The Relation between Risk
and Level of Chemical Components
in Drinking Water

Chemical analysis of drinking water
in Khartoum State

Eiman M. Ibrahim

2021

First Printing: 2021

ISBN 978-1-329-63859-4

Any websites referred to in this publication are in the public domain, and their addresses are provided for information only. The publishers and editors disclaim any responsibility for the content.

Euro-Khaleeji Research and Publishing House
Sultanate of Oman
www.euro-khaleeji.org

The cover image is by Katja Just from Pixabay.

In the beginning

قال الله تعالى:

﴿يَرْفَعِ اللَّهُ الَّذِينَ آمَنُوا مِنكُمْ وَالَّذِينَ أُوتُوا الْعِلْمَ دَرَجَاتٍ وَاللَّهُ بِمَا تَعْمَلُونَ خَبِيرٌ﴾

صدق الله العظيم

سورة المجادلة الأية ﴿١١﴾

"Allah will elevate those of you who are faithful and raise those gifted with knowledge in rank. And Allah is All-Aware of what you do."

Dedication

I dedicate this book to express my heartiest gratitude to the respected people who supported me from the very beginning.

The particular people who deserve my special appreciation are as follows:

The College of Environmental Studies and Disaster Management, The National Ribat University, Khartoum State Water Corporation-Central Water Lab, UNESCO Chair in Water Resource-Water Quality Lab, and Mr. Babiker Siddig Khartoum University.

I take this opportunity to extend my gratitude to the International Education Consultancy - IEC. The completion of this book could not have been a reality without its services.

I express my thanks to ICEM College.

I dedicate this book to all patients who have been affected by chemical pollutions.

I am thankful to my family.

Thanks to Allah, who enabled me to write a book about such an important topic.

Abstract

Overall goal

This study's overall goal is to assess water quality from wells by chemical analysis of 23 water samples taken from different areas in Khartoum, Omdurman, and Bahri and compare the results with WHO SSMO standards.

Methodology

This study utilized spectrophotometry, Test-tube Lab Research for chemical analysis, and a cross-sectional study design were based on questionnaires.

Results

The analysis results of water samples were collected from 2012 to 2017 from different water sources: Groundwater and wells water from surface water sources such as Blue Nile, White Nile, and River Nile (and treated and untreated) from Khartoum State. The study concentrated on conducting a chemical analysis on water samples taken from wells near the industrial areas in Khartoum state.

The parameters of chemical and physical analysis were taken from experimental methods, such as Electric conductivity, suspended solids, Temperature, Turbidity, Total dissolved solids (TDS), and all these parameters indirect measurement reading using Standard (method NO 2130). Absorption Spectrophotometer (Model 1100) and the titration for some ions, and SO_4^{-2}, Cl^-, HCO_3^{-2}, and CO_3^{-2} were analyzed using a Dionex ion, chromatograph

2000i, and analyses ionic components that included: Magnesium, Calcium, Free Iron total, hardness(total), Total alkalinity, Nitrate, Nitrite, Ammonia, Hydrogen sulfide & sulfides, Sulfate and Iron.

The Study Design utilized a cross-sectional study and was based on questionnaires. The data which were collected were statistically analyzed using a Statistical Program for Social Science (SPSS). The results of all samples' parameters were according to the WHO standards for the Khartoum area, except for some data. These are as follows: turbidity, the odor test, which showed positive results indicating that the water was polluted, Ammonia and dissolved iron concentrations were more than the WHO and SSMO standards for samples from some locations.

Recommendations

To improve the quality of water in Sudan, this study recommends:

1. Increase the awareness of the community about the effects of chemical pollution of water.

2. The government should provide all citizens with safe water meeting the WHO standards.

3. Use Isotopes Scopes to analyze and purify water in Khartoum states and other states in the entire labs.

4. Use solar energy for operating machines in stations for purifying drinking water in Khartoum state and other states.

5. The Government should invest in the water industry to be provided with safe drinking water.

6. In each residential block, there should be a network of sewage treatment.

7. Sudan should implement the millennium Development Goal (M.D.GS), which Sudan had ratified.

8. Periodical analysis for drinking water wells is critical.

9. Conduct scientific research on the quantities and qualities of groundwater in Khartoum state.

10. Construct dams to store rains water to be used after treatment in Khartoum state's rural areas.

Table of Contents

ABSTRACT..IV

OVERALL GOAL IV

METHODOLOGY.. IV

RESULTS IV

RECOMMENDATIONS VI

TABLE OF CONTENTS...................................VII

ABBREVIATIONS..XI

1. CONTEXTUAL BACKGROUND........................1

1.1 BACKGROUND...1

1.2 PROBLEM IDENTIFICATION 3

1.3 JUSTIFICATION 4

1.4 HYPOTHESIS OF WATER CHEMICAL POLLUTION: 4

1.5 OBJECTIVES.. 5

1.5.1 Overall objective 5

1.5.2 Specific objectives 5

2. SOURCES IN THE FIELD 7

2.1 POPULATION GROWTH AND URBAN EXTENSION OF GREATER KHARTOUM.. 7

2.2 WATER PRODUCTION 9

2.3 THE PRINCIPAL PARAMETERS OF WATER QUALITY ACCORDING TO WHO.................................... 12

1. Acidity 12

2. Alkalinity 13

3. Ammonia 15

4. Calcium.. 17

5. Chloride 17

6. Conductivity.....................................18

7. Fluoride 19

8. Iron 20

9. Magnesium 21

10. Manganese 22

11. Nitrate 22

12. Nitrite 24

13. pH...25

14. Phosphates.................................... 26

15. Potassium 27

16. Sodium...27

17. Sulphate.. 28

18. Temperature................................... 29

2.4 GROUNDWATER AND PUBLIC HEALTH 30

2.4.1 Effects of Ground Water Pollution 30

2.4.2 Effects of drinking water on human health in Sudan...32

2.5 GROUNDWATER AS A SOURCE OF DRINKING WATER 34

2.6 DISEASE DERIVED FROM GROUNDWATER USE................ 36

2.6.1 Chemical hazards... 36

2.6.2 Infectious disease transmission 37

3. MATERIAL AND METHODS 39

3.1 MATERIALS 39

3.2 METHODOLOGY 39

3.2.1 Data of the questionnaire................................ 39

1. Study Area 40

2. Experimental work 41

3. Study Area and Duration 41

4. Chemical tests... 43

4. RESULT.. **49**

5. DISCUSSION, CONCLUSION, AND RECOMMENDATIONS
.... **74**

5.1. DISCUSSION 74
    1. Analysis results of the questionnaire 74
    2. Analysis of the results of water samples 76
5.2. CONCLUSION 82
5.3. RECOMMENDATIONS 84

REFERENCES.. **85**

APPENDICES **88**

FIGURE 1: THE MEAN OF PARAMETERS IN STUDY AREAS 88
QUESTIONNAIRE... 89
RESULTS 93
    Location: 1.. 93
    Location: 2 93
    Location: 3 94
    Location: 4 94
    Location: 5 95
    Location: 6 95
    Location: 7 96
    Location: 8 96
    Location: 9 97
    Location: 10.. 97
    Location: 11 98
    Location: 12.. 98
    Location: 13.. 99
    Location: 14.. 99
    Location: 15... 100
    Location: 16... 100

Location: 17 101
Location: 18 101
Location: 19 101
Location: 20 102
Location: 21 102
Location: 22 103
Location: 23 103

Abbreviations

Abbreviation	Meaning
WHO	World Health Organization
WFD	Water Framework Directive
CWA	Clean Water Act
HUC	Hydrologic Unit Code
TDS	Total Dissolved Solids
GW	Ground Water
SPSS	Statistic Program for Social Science Statistics
MDGs	Millennium Development Goals
USGS	The United States Environmental Protection Agency
WWC	World Water Council
BGS	British Geological Survey
DPHE	Department of Public Health and Environment

1. Contextual Background

1.1 Background

Water is considered a necessary resource and is the elixir of life. Water is composed of about 70% of the bodyweight of almost all living organisms. Without water, life is not possible on this planet. Water, of a natural origin which has been utilized for various objectives, mainly for drinking, domestic, irrigation, and industrial, relies on its intrinsic quality; hence it is of primary importance to have preliminary information on the quality and quality of water resources obtainable in the region. It is expected that numerous people worldwide may suffer from water shortage problems and suffer from the acute water crisis.

There are many water sources in Sudan. In particular, surface water resources are examples of such sources. On the other hand, groundwater is only restricted to small areas. The biggest supplier of surface water is the Nile River. This river has a significant part where three streams flow together to create this large river. The tributaries of the substantial portion of the Nile River are as follows, the Blue Nile (65%) and the White Nile (23%) in the capital Khartoum, and the Atbara River (12%). The Atbara and the Blue Nile rivers emerge from the Atbara and the Blue Rivers' Ethiopian plateau. However, the Equatorial Lakes Plateau is the main area where the White River starts to flow. Besides, Gash and Baraka's seasonal rivers are found in eastern Sudan. Niles and multiple boreholes drilled are the primary sources that supply water for Greater Khartoum (NATIONS, 1995).

The River Nile and its streams are the main sources that provide water in Sudan. According to (Abdeen, 2017), Sudan has an annual rainfall estimated at 1093.2 x 109 m3. The northern desert and the tropical region are known to have annually rainwater ranging from 1 mm to about 1600 mm. When circumstances change and rainwater dries out, utilizing open wells as the water source is an efficient solution. They can be made by digging land in rural areas so that humans and animals can be supplied by water.

Besides all these natural sources of water, the demand for groundwater becomes a significant need.

Diseases that are related to water are one of the leading health issues worldwide. Examples of such diseases are diarrheal diseases. They caused 1.8 million deaths in 2002 and appeared to represent around 62 million Disability Adjusted Life Years per annum. This makes diarrheal diseases the sixth highest cause of mortality and the third-highest of morbidity. It is considered to represent 3.7 percent of the global disease burden, and it is consequent from low water, sanitation, and hygiene. In 2002, 1.1 billion people did not contain improved water supply within 1 kilometer of their homes, and about 2.6 billion people could not find some form of enhanced excreta disposal. (Oliver Schmoll, 2006).

It was revealed in Sudan during the past 50 years that the quality of drinking water possessed terrible effects on human bodies as some symptoms of epidemiological disease occurred. Examples of such diseases are cardiovascular mortality and atherosclerosis urolithiasis, dental caries, dental fluorosis, and kidney failure. Moreover, there was some correlation between the natural mineral-contents of the drinking water, for example, TDS, TH, F$^-$ and NO3$^-$, and the occurrence of these diseases. Hardness in water can be defined as having multivalent cations. Furthermore, atherosclerosis's clinical occurrence is considered to be higher in soft-water areas than in hard-water areas. Some studies relate water's softness to ischemic heart disease, hypertension, atherosclerosis, and stroke (Abdellah, Abdel-Magid, & Shommo, Assessment of Groundwater Quality in Southern Suburb of the Omdurman City of Sudan, 2013).

1.2 Problem Identification

Water is the backbone of life; it regulates vegetation and animals' distribution, determines the kind of climate, and affects the human environments. This can be the explanation that a lot of ancient civilizations developed in the Well water. Wells and streams are internationally the major features of most types of landscape. They play a crucial role in socio-economic progress. In Africa, they can be regarded as the key to development.

The causes behind the poor awareness of the community's individuals in water pollution by chemicals and the diseases caused by this pollution in Sudan remain unclear. Whether this unwillingness is a result of personal, educational, or economic issues needs to be determined.

This study intends to improve the community's awareness about chemical water pollution after getting the experimental methods of chemical water analysis to study the environment and its relation to the pollution of drinking water and prevent people from the diseases caused by this type of pollution.

3

1.3 Justification

1. The study can create an intelligent system for water-quality analysis based on subjectivity and uncertainty in the assessment.

2. The study can lead to a high level of understanding of water chemical pollution; there will be more clarification on how to increase public awareness and educate people about pollution in the environment.

3. To highlight the importance of water quality issues is of high priority and included in policies and adequately funded funding. To advocate for legislation, awareness and extension are low fields in water quality management.

1.4 Hypothesis of water chemical pollution:

• Industrial and agricultural work includes numerous different chemicals that can run-off into the water and pollute it.

• Can contaminate rivers and lakes by throwing metals and solvents during industrial work. Those metals and solvents possess toxicity to numerous aquatic organisms and

therefore cause defects in their development or, consequently, death.

The analysis of some metals, such as sodium and magnesium, does not meet the WHO parameters.

• Pesticides are defined as chemicals that control weeds, insects, and fungi that existed in farms. As a result of using pesticides, there will be a higher risk of causing water pollution and poisoning aquatic organisms. Consequently, if birds, humans, and animals feed on infected fish, they will poison such organisms.

• Petroleum, cement industry, and mining such as gold, are considered as chemical pollutants.

1.5 Objectives

1.5.1 Overall objective

Develop a conceptual model to assess water quality in wells from a perspective of environmental risk assessment, including a comprehensive way to manage linguistic uncertainty and subjectivity.

1.5.2 Specific objectives

1. To gain a greater understanding of water chemical pollution, understanding how pollution travels and persists in the environment can play an important part in public awareness, education, and applying political pressure. Good science can inform policy and legislation and empower campaigners, but science alone rarely makes much difference.

2. To examine the nature of technical change and its consequence on pollution generation over some time

3. To detect the status of pollution control through different abatement techniques

4. To determine the pH and temperature optima for water

5. To create an intelligent system for water-quality analysis based on subjectivity and uncertainty in the assessment

6. To propose an integrated approach to deal with multi-chemical screening risk assessment in river sediments.

2. Sources in the field

People can survive days, weeks, or months with no food, however only about four days without water. Water, though an absolute requirement for life, can be a transporter of many diseases. Water can be hard or soft, natural or modified, bottled, or tap.

Sudan is considered one of Africa's largest countries and is found mainly in the arid region, where water is rare. Water used in Sudan and the Nile basin is derived almost exclusively from surface water resources, as groundwater is used in other areas.

Furthermore, as groundwater is rich in Sudan, this has made the demand for groundwater grows and has led the government to drill more wells (NATIONS, 1995).

Water quality is a term used to explain the appropriateness of water to sustain various uses or processes. Water quality can be described by a range of variables that serve as a limit to water use (Bartram & Ballance, 1996).

2.1 Population growth and urban extension of Greater Khartoum

The number of population in Greater Khartoum showed a major rise following the indepence of South Sudan in the of 1990s. Consquently, the transportation and manufacturing industries were advanced (Beckedorf, 2012).

Different sources estimate the following numbers, according to which the number of inhabitants of Greater Khartoum increased almost eight folds in 38 years, as shown in the following table:

Table 1: Estimated population growth in Greater Khartoum, 1990 to 2000[1]

Area	Year 1990	Year 2000
Khartoum Centre	50.000	100.000
Khartoum South-West	750.000	1.200.000
Khartoum South-East	350.000	1.200.000
Total	1.150.000	2.500.000
North-Bahri	250.000	350.000
East-Bahri	550.000	1.100.000
Total	800.000	1.450.000
Omdurman North	450.000	550.000
Omdurman Centre	250.000	400.000
Omdurman South-East	650.000	800.000
Total	1.350.000	1.750.000
Khartoum Total	3.300.000	5.300.000

Table 2: Estimated population growth in Khartoum, 1913 to 2008 [2]

---

[1] Ref: Central Bureau of Statistics https://www.cbs.gov.sd
https://www.populationpyramid.net/ar/suda /2017/
[2] Ref: Central Bureau of Statistics
https://www.cbs.gov.sdhttps://www.populationpyramid.net/ar/sudan /2017/

The following table shows treatment plants and pumping stations in Khartoum from 1 long with the accelerated population growth, Greater Khartoum expanded in the area as well as delivered in the table (2)

Year	Inhabitants of Greater Khartoum
1913	110.000
1956	260.000
1966	460.000
1973	780.000
1983	1.340.000
1993	3.800.000
2005	4.700.000
2008	5.300.000
2009	551.525.7
2010	575.823.4
2011	600.688.5
2012	626.793.0
2013	653.479.5
2014	680.904.6
2015	709.514.8
2016	738.515.8
2017	768.754.7

2.2 Water production

There are three Niles that Khartoum State, Sudan relies on. These Niles are the Nile River's fresh surface water, the Blue Nile, and the White Nile. All these Niles reach the Mugran area. For treating surface water, pipelines are needed. Then, this surface water will be circulated to citizens and residents. In Khartoum area, exploring groundwater has been considered one of the main issues that the governmental and private sectors focus on (Ahmed, Sulaiman, Osman, Saeed, & Mohamed, 2000).

The water supply system in Sudan (mainly Khartoum) has faced numerous issues that occurred due to the higher desire of people to get drinking water. Citizens of Sudan can get their water supply from two sources. These sources are groundwater and river water from the Nile. Wells are used to get groundwater. Whereas, treatment plants are used to get river water from the Nile (D. Stephenson, 2004).

The treatment plants in Khartoum underwent multiple developmental steps. Burri and Betelmal were the first areas to contain water treatment plants. These two treatment plants were built in 1924 (Burri) and 1927 (Betelmal). According to the timeline, the following treatments plants that were established after the first plants were Old Bahri (1954), New Bahri (1979), Moghran (1964), and Tuti Island (1984). During the 2000s, there were four new treatment plants in Soba, GebelAulia, Shamal Bahri, and ElManara . These treatment plants were established in 2009 and 2010 (Beckedorf, 2012).

Table 3: Treatment plants and pumping stations in Khartoum, 2010 [3]

Year	Main financier	Name	Production (in cubic meter per day)
1924	Britain	Burri	16.000 extension to 50.000 under construction

[3] Ref: Science & Engineering Journal, Vol.6, Issue 1, 2013. Introducing and Implementing an EMS in Khartoum State Water Corporation Nile Basin water.

1927	Britain	Betelmal	25.000 since 2003 (formerly 12.000)
1954	Britain	Old Bahri	12.000; merged with New Bahri in 1979
1964	Britain & France	Moghran	90.000 since 1990
1979	Czechoslovakia	New Bahri	180.000 since 1999 (formerly 90.000); extension to 300.000 under construction
1984	Sudan	Tuti	2.000
1992	Japan	El-Gomayir	Pumping Station Storage of 50.000 since 2010(formerly 36.000
2002	Egypt	Sahafa	Pumping Storage of 80.000 since 2006 (formerly 63.000) Station
2009	Egypt	Soba	100.000 extension to 200.000 under construction
2010	Iran	Gebel Aulia	68.000 production of 30.000 in 2010 due to lack of Networks
2010	Spain	Nord Bahri	50.000 production of 20.000 in 2010 due to intake problems
2010	Britain	El Manara	200.000 capacity production of 50.000 in 2010
Under Construction	Spain	Aid Babikir	Pumping Station 50.000 projected
Under Construction	Iran	Abu Said	200.000 projected

Table 4: Wells drilled in Greater Khartoum, 1960 to 2014[4]

Decade	Khartoum	Bahri	Omdurman	Total
1961 to 1970	26	26	46	98
1971 to 1980	54	41	57	152
1981 to 1990	23	16	39	78
1991 to 2000	44	58	78	180
Total	147	141	220	508

The number of the regular wells and the high productivity wells that were registered in 2010 were 1300 and 60 respectively. It is determined that the time needed to process one well ranges between five and ten hours (Beckedorf, 2012). Khartoum state is currently supplied about 52% by groundwater and 48% by water extracted

---

[4] Ref: Science & Engineering Journal, Vol.6, Issue 1, 2013. Introducing and Implementing an EMS in Khartoum State Water Corporation Nile Basin Water.

from the Nile River. Approximately more than 500.000 cubic meters of water per day is produced in Khartoum [5].

Based on (Mohamed & Skerratt, 2013), it is stated that there are three cities that contain the treatment plant's water network. These cities are Khartoum, Bahri, and Omdurman. These areas involve treated water that is received from the Nile River. This water is considered a combination between water from treatment plants and groundwater wells. Furthermore, there are cities that are riched with this hybrid water network, and the reason is that they are close to the main central districts. These areas are El Thoura all the way to to the north of Omdurman. Also, Umbadda to the south of Omdurman, or Sahafa to the south of Khartoum. Wells of groundwater nourish and supply the local well networks only, which are found in Dar Essalam in the west of Omdurman or Hag Youssef in the east of Bahri. Moreover, donkey carts can be utilized to provide water in areas where water supply by networks is unreliable due to numerous water cuts and pipeline breakdowns.

2.3 The Principal Parameters of Water Quality according to WHO

1. Acidity

Chemical Symbol or Formula: Not Applicable [Bulk parameter].

Units Used for Analytical Results: mg/l $CaCO_3$.

---

[5] Ref: Science & Engineering Journal, Vol.6, Issue 1, 2013. Introducing and Implementing an EMS in Khartoum State Water Corporation Nile Basin Water.

Standard Method(s) of Analysis: Titration with Sodium Hydroxide [A]

Occurrence/Origin: Arises from the presence of weak or strong acids and/or certain inorganic salts. The dissolved carbon dioxide presence is usually the main acidity factor in unpolluted surface waters (it forms the weak acid H_2CO_3 - carbonic acid).

Health/Sanitary Significance: No particular implications apart from palatability considerations in excessively acid waters.

Background Information: The determination is arbitrary to a degree in that the end-point pH values [color changes] will depend on the choice of indicator reagents made by the analyst.

Commonly, methyl orange or bromophenol blue are used in the first stage as the indicator (color change around pH 4.5), with phenolphthalein in the second stage (color change around pH 8.3).

Hence the various terms methyl orange acidity (alternatively, mineral acidity) and phenolphthalein acidity (or total acidity). (WHO, 2001)

2. Alkalinity

Chemical Symbol or Formula: Not Applicable [Bulk parameter].

Units Used for Analytical Results: mg/l $CaCO_3$.

Standard Method(s) of Analysis: Titration with Sulphuric Acid [A].

Occurrence/Origin: The alkalinity of natural water is generally due to the presence of bicarbonates formed in reactions in the soils through which the water percolates. It measures the water's capacity to neutralize acids and reflects its so-called *buffer capacity* (its inherent resistance to pH change). Poorly-buffered water will have low or very low alkalinity and will be susceptible to pH reduction by, for example, "acid rain." At times, however, river alkalinity values of up to 400 mg/l $CaCO_3$ may be found; they are without significance in the water quality.

Health/Sanitary Significance: There is little known sanitary significance attaching to alkalinity (even up to 400 mg/l $CaCO_3$), though unpalatability may result in highly alkaline water.

Background Information: Alkalinity in natural waters may also be attributable to carbonates and hydroxides. Sometimes analysis is carried out to distinguish between the alkalinity elements, and this is done by using different indicators in the titration procedure and making appropriate calculations. The indicators most commonly employed are phenolphthalein (color change around pH8.3) and methyl orange (color change around pH4.5), resulting in the additional terms phenolphthalein alkalinity and methyl orange alkalinity; the latter is synonymous with total alkalinity.

Alkalinity is involved in the consequential effects of eutrophication [over-enrichment] of waters.

Where a high degree of photosynthesis occurs, as discussed below under "Oxygen, Dissolved" (q.v.), there is high consumption of carbon dioxide by algae. As any free carbon dioxide initially available is consumed, more is produced in a series of related chemical equilibrium reactions, as follows:

1. $H^+ + HCO_3^{-2} \Leftrightarrow H_2CO_3 [H_2O + CO_2]$

2. $H^+ + CO_3^{-2} \Leftrightarrow HCO_3^-$

3. $H_2O \Leftrightarrow H^+ + OH^-$.

As the carbon dioxide is consumed by photosynthesis, more is produced (reaction 1, left to right) by bicarbonate ions, present as alkalinity, and hydrogen ions to give undissociated carbonic acid (carbon dioxide and water). Any carbonate ions present will react with more hydrogen ions to replace the bicarbonate consumed (reaction 2, again left to right). These reactions consume hydrogen ions, which are produced as reaction 3 (equilibrium again to the right). A net overall effect is the production of hydroxyl ions and an increase in the pH. It is not uncommon for an extreme photosynthetic activity to produce pH levels high enough to cause severe damage (even death) to fish (WHO, 2001).

3. Ammonia

Chemical Symbol or Formula: NH_3.

Units Used for Analytical Results: mg/l N.

Standard Method(s) of Analysis: Colorimetric (Manual; Nessler's Reagent) [A/B]; Colorimetric (Automated; Berthelot Reaction) [B/C].

Occurrence/Origin: Ammonia is generally present in natural waters, though in tiny amounts, due to microbiological activity, which causes the reduction of nitrogen-containing compounds. When present in levels

15

above 0.1 mg/l N, sewage or industrial contamination may be indicated.

Health/Sanitary Significance: From the viewpoint of human health, the significance of ammonia is marked because it indicates the possibility of sewage pollution and the consequent possible presence of pathogenic micro-organisms.

Background Information: The form of the ammonia - whether it is "free" (as NH_3) or "saline" (as NH_4^+) in slightly acid waters - depends on the pH, and these forms are not distinguished from one another during analysis. The different terms commonly applied to the forms of ammonia are as follows: total ammonia (NH_3 & NH_4^+), total ammonium (NH_3 & NH_4^+) free ammonia (NH_3) free and saline ammonia (NH_3 & NH_4^+), ionized ammonia (NH_4^+), and un-ionized ammonia (NH_3).

The different forms arise from the pH/temperature related equilibrium reactions:

$$1.\ NH_3 + H_2O \Leftrightarrow NH_3.H_2O$$

$$2.\ NH_3.H_2O \Leftrightarrow NH_4^+ + OH^-$$

The ammonia tolerances for fishery waters are narrow and have been considered and reported on by the European Inland Fisheries Advisory Commission. It is the un-ionized ammonia species that is most harmful to freshwater aquatic life and game fish. Arising from the complex relationship between total ammonia concentration, pH, and temperature, there emerges a level for total ammonia of around 0.3 mg/l NH_3, which is considered to be that which would contain the limiting amount of un-ionized ammonia (WHO, 2001).

4. Calcium

Chemical Symbol or Formula: Ca.

Units Used for Analytical Results: mg/l Ca.

Standard Method(s) of Analysis: Titration (Calcium Hardness) [A]; Atomic Absorption Spectrometry [B].

Occurrence/Origin: Occurs in rocks, bones, shells, etc. Very abundant.

Health/Sanitary Significance: High levels may be beneficial, and waters that are rich in calcium (and hence are tough) are very palatable. (WHO, 2001)

5. Chloride

Chemical Symbol or Formula: Cl^-

Units Used for Analytical Results: mg/l Cl.

Standard Method (s) of Analysis: Titration (Mohr Method: Silver Nitrate) [A].

Occurrence/Origin: Chloride exists in all natural waters, the concentrations varying very widely and reaching a maximum in seawater (up to 35,000 mg/l Cl).

In freshwaters, the sources include soil and rock formations, sea spray, and waste discharges. Sewage contains large amounts of chloride, as do some industrial effluents.

Health/Sanitary Significance: Chloride does not pose a health hazard to humans, and the principal consideration is palatability.

Background Information: At levels above 250 mg/l, Cl water will begin to taste salty and become increasingly objectionable as the concentration rises further.

However, external circumstances govern acceptability, and in some arid areas, waters containing up to 2,000 mg/l Cl are consumed, though not by people unfamiliar with such concentrations. High chloride levels may similarly render freshwater unsuitable for agricultural irrigation.

Comments: Because sewage is such a rich source of chloride, a high result may indicate water pollution by a sewage effluent. Natural levels in rivers and other freshwaters are usually in the range 15-35 mg/l Cl⁻ much below drinking water standards. What is normally important to note in a series of results from a river, for example, is not the absolute level but rather the relative levels from one sampling point to another.

An increase of even 5 mg/l at one station may increase suspicions of a sewage discharge, especially if the free ammonia levels (q.v.) are also elevated. However, in coastal areas, elevated chloride values may be due to sea spray or seawater infiltration, not necessarily discharges. Normal raw water treatment processes do not remove chloride (WHO, 2001).

6. Conductivity

Chemical Symbol or Formula: Not Applicable [Physical parameter].

Units Used for Analytical Results: μS/cm.6

Standard Method(s) of Analysis: Electrometric [A].

Occurrence/Origin: Reflects the mineral salt content of water.

Health/Sanitary Significance: No direct significance.

Background Information: Also referred to as electrical conductivity and, not wholly accurately, as specific conductance, the conductivity of water is an expression of its ability to conduct an electric current.

As this property is related to the ionic content of the sample, which is, in turn, a function of the dissolved (ionizable) solids concentration, the relevance of easily performed conductivity measurements is apparent.

In itself, conductivity is a property of little interest to a water analyst. Still, it is an invaluable indicator of the range into which hardness and alkalinity values are likely to fall, and also in the order of the dissolved solids content of the water.

While a certain proportion of the dissolved solids (for example, those which are of vegetable origin) will not be ionized (and hence will not be reflected in the conductivity figures) for many surface drinking of water, the following approximation will apply: Conductivity (μS/cm) x 2/3 = Total Dissolved Solids (mg/l) (WHO, 2001).

7. Fluoride

Chemical Symbol or Formula: F^-.

Units Used for Analytical Results: mg/l F.

Standard Method(s) of Analysis: Colorimetric (after distillation) [B]; Specific Ion Electrode [B].

Occurrence/Origin Occurs naturally in quite rare instances; it arises almost exclusively from fluoridation of public water supplies and industrial discharges.

Health/Sanitary Significance: Health studies have shown that the addition of fluoride to water supplies in levels above 0.6 mg/l F leads to a reduction in tooth decay in growing children and that the optimum beneficial effect occurs around 1.0 mg/l.

Background Information: At levels markedly over 1.5 mg/l, an inverse effect occurs, and mottling of teeth (or severe damage at gross levels) will arise. For this reason, there is a constraint on fluoride levels, the effects of which vary with temperature (WHO, 2001).

8. Iron

Chemical Symbol or Formula: Fe.

Units Used for Analytical Results: mg/l Fe.

Standard Method(s) of Analysis: Colorimetric (o-Phenanthroline) [B]; Atomic Absorption Spectrometry [B/C].

Occurrence/Origin: Geological formations (especially under reducing conditions); acid drainage; effluent discharges.

Health/Sanitary Significance: The objections to iron are primarily organoleptic, but there have been recent medical concerns about high drinking water levels.

Background Information: Iron is present in significant amounts in soils and rocks, principally in insoluble forms. However, many complex reactions that occur naturally in ground formations can give rise to more soluble forms of iron, resulting in water passing through such formations.

Appreciable amounts of iron may therefore be present in groundwaters.

Comments: The metal is quite harmful to aquatic life, as evidenced by laboratory studies. In nature, the degree of toxicity may be lessened by the iron's interaction with other water constituents.

The metal is converted to an insoluble form, and then the iron deposits will interfere with fish food and spawning (WHO, 2001).

9. Magnesium

Chemical Symbol or Formula: Mg.

Units Used for Analytical Results: mg/l Mg.

Standard Method(s) of Analysis: Titration with EDTA [A]; Atomic Absorption Spectrometry [B/C].

Occurrence/Origin: Major constituent of geological formations.

Health/Sanitary Significance: Indirect (in conjunction with Sulphate, q.v.).

Background Information: Like calcium (q.v.), magnesium is abundant and a major dietary requirement for humans (0.3-0.5 g/day).

It is the second major constituent of hardness, and it generally comprises 15-20 percent of the total hardness expressed as $CaCO_3$.

21

Its concentration is very significant when considered in conjunction with that of sulfate (q.v.).

Comments: Magnesium sulfate is used medicinally as "Epsom Salts," a laxative.

The decreasing hardness of water caused by magnesium can occur by reacting it with the chemical compound under WHO standards (WHO, 2001).

10. Manganese

Chemical Symbol or Formula: Mn.

Units Used for Analytical Results: mg/l Mn.

Standard Method(s) of Analysis: Colorimetric (Persulphate) [B]; Atomic Absorption Spectrometry [B/C].

Occurrence/Origin: Widely distributed constituent of ores and rocks.

Health/Sanitary Significance: No particular toxicological connotations; the objections to manganese - like iron - are aesthetic.

Background Information: As with iron, manganese is found widely in soils and constitutes much groundwater. It may be brought into solution in reducing conditions, and the excess will be later deposited as the water is reaerated. The general remarks for iron (q.v.) apply to manganese, but the staining problems with this metal may be even more severe, hence the quite stringent limits. A second effect of the presence of manganese much above the limits is an unacceptable taste problem (WHO, 2001).

11. Nitrate

Chemical Symbol or Formula: NO_3^-

Units Used for Analytical Results: mg/l N or mg/l NO_3^-.

Standard Methods of Analysis: Manual/Automated Colorimetry [A/B].

Occurrence/Origin: Oxidation of ammonia: agricultural fertilizer run-off.

Health/Sanitary Significance: Hazard to infants above 11 mg/l N [50 mg/l NO_3^-].

Background Information: Relatively little of the nitrate found in natural waters is of mineral origin, most coming from organic and inorganic sources, the former including waste discharges and the latter comprising chiefly artificial fertilizers. However, bacterial oxidation and fixing of nitrogen by plants can both produce nitrate. Interest is centered on nitrate concentrations for various reasons.

Most importantly, high nitrate levels in waters to be used for drinking will render them hazardous to infants as they induce the "blue baby" syndrome (methemoglobinemia).

The nitrate itself is not a direct toxicant. Still, it is a health hazard because of its conversion to nitrite, which reacts with blood hemoglobin to cause methemoglobinemia.

Sewage is rich in nitrogenous matter, which through bacterial action may ultimately appear in the aquatic environment as nitrate.

Hence, the presence of nitrate in groundwaters, for example, is cause for suspicion of past sewage pollution or excess levels of fertilizers or manure slurries spread on land.

(High nitrite levels would indicate more recent pollution as nitrite is an intermediate stage in the ammonia-to-nitrate oxidation).

In rivers, high nitrate levels are more likely to indicate significant run-off from agricultural land than anything else. However, it should be noted that there supply. Nitrite concentrations in rivers are rarely more than 1 - 2 percent of the nitrate level. Therefore, it may be acceptable to carry out the analytically convenient determination of nitrate + nitrite at the same time. This determination is correctly referred to as total oxidized nitrogen (WHO,2001).

12. Nitrite

Chemical Symbol or Formula: NO_2^-

Units Used for Analytical Results: mg/l NO_2-.

Standard Methods of Analysis: Manual or Automated Colorimetry [A/B]

Occurrence/Origin: Generally, from untreated or partially treated wastes.

Health/Sanitary Significance: Methaemoglobinaemia-causing agent [cf. Nitrate].

Background Information: Nitrite naturally exists in low concentrations, and even in waste treatment, plant effluents levels are relatively low, principally because the nitrogen will tend to exist in the more reduced (ammonia; NH_3) or more oxidized (nitrate; NO_3) forms.

Because nitrite is an intermediate in ammonia's oxidization to nitrate, such oxidation can proceed in soil.

Because sewage is a rich source of ammonia nitrogen, waters that show any appreciable amounts of nitrite are regarded as being of highly questionable quality. Levels in unpolluted waters are generally low, below 0.03 mg/l NC2. Values greater than this may indicate sewage pollution (WHO, 2001).

13. pH

Chemical Symbol: Not applicable [Physical parameter].

Units Used for Analytical Results: pH units.

Standard Method(s) of Analysis: Electrometry [pH electrode] [A/B]

Occurrence/Origin: Physical characteristic of all waters/solutions.

Health/Sanitary Significance: None - except that extreme values will show excessive acidity/alkalinity, with organoleptic consequences.

Background Information: By definition, pH is the negative logarithm of a solution's hydrogen ion concentration. It is thus a measure of whether the liquid is acid or alkaline.

The pH scale (derived from the ionization constant of water) ranges from 0 (very acid) to 14 (very alkaline). The range of natural pH in freshwaters extends from around 4.5 for acid, peaty upland waters to over 10.0 in waters with intense photosynthetic activity by algae. However, the most frequently encountered range is 6.5-8.0.

In waters with low dissolved solids, which consequently have a low buffering capacity (i.e., low internal resistance to pH change), changes in pH induced by external causes may be quite dramatic.

Extremes of pH can affect a water's palatability, but the corrosive effect on distribution systems is a more urgent problem.

The effect of pH on fish is also an important consideration. Values that depart increasingly from the normally found levels will have a more marked impact on fish, leading ultimately to mortality. The range of pH suitable for fisheries is 5.0-9.0, though 6.5-8.5 is preferable (WHO, 2001).

14. Phosphates

Chemical Symbol or Formula: PO_4^{---}.

Units Used for Analytical Results: mg/l P.

Standard Method (s) of Analysis: Manual or Automated Colorimetry [B/C].

Occurrence/Origin: Phosphorus occurs widely in plants, in microorganisms, in animal wastes, and so on. It is widely used as an agricultural fertilizer and a principal constituent of detergents, mainly for domestic use. Run-off and sewage discharges are thus important contributors of phosphorus to surface waters.

Health/Sanitary Significance: None.

Background Information: The significance of phosphorus is principal regarding the phenomenon of eutrophication (over-enrichment) of lakes and, to a lesser extent, rivers.

Phosphorus gaining access to such water bodies, along with nitrogen as nitrate, promotes the growth of algae and other plants, leading to blooms, littoral slimes, diurnal dissolved oxygen variations of great magnitude, and related problems, as discussed elsewhere in this volume (WHO, 2001).

15. Potassium

Chemical Symbol or Formula: K.

Units Used for Analytical Results: mg/l K.

Standard Method(s) of Analysis: Flame Photometry [B/C], Atomic Absorption Spectrometry [B/C].

Occurrence/Origin: Geological formations.

Health/Sanitary Significance: None, except at gross levels.

Background Information: Potassium is an essential constituent of many artificial fertilizer formulations. Hence, its determination is often carried out on lake waters when an assessment is being made of nutrient input. However, potassium tends to be "fixed" in soils and is not that easily leached out.

There are no implications of toxicity (WHO, 2001).

16. Sodium

Chemical Symbol or Formula: Na.

Units Used for Analytical Results: mg/l Na.

Standard Method(s) of Analysis: Flame Photometry [B/C]; Atomic Absorption Spectrometry [B/C].

Occurrence/Origin: Abundant constituent of rocks and soils.

Health/Sanitary Significance: Causes hypertension if taken in excess.

Background Information: Sodium is always present in natural waters. It is also an essential dietary requirement, and the regular intake is as common salt (sodium chloride) in food; daily consumption may amount to 5 grams or more. The main reason for limiting it is the combined effect, which exercises with sulfate (see below). Still, too excessive an intake (the latter normally being 2-3 times the dietary threshold) can cause hypertension, as mentioned (WHO, 2001).

17. Sulphate

Chemical Symbol or Formula: $SO4^-$

Units Used for Analytical Results: mg/l SO_4^-

Standard Method(s) of Analysis: Turbid metric (Barium Sulphate) [B/G]; Ion Chromatography[C].

Occurrence/Origin: Rocks, geological formations, discharges, and so on.

Health/Sanitary Significance: Excess sulfate has a laxative effect, especially in combination with magnesium and/or sodium.

Background Information: Sulphates exist in nearly all natural waters, the concentrations varying according to the nature of the terrain through which they flow. They are often derived from heavy metals' sulfides (iron, nickel, copper, and lead). Iron sulfides are present in sedimentary rocks. They can be oxidized to sulfate in humid climates; the latter may then leach into watercourses so that groundwaters are often excessively high in sulfates.

As magnesium and sodium are present in water, their combination with sulfate will have an enhanced laxative effect of greater or lesser magnitude depending on concentration. Therefore, the utility of water for domestic purposes will be severely limited by high sulfate concentrations, hence the limit of 250 mg/l SO_4^- (WHO, 2001).

18. Temperature

Chemical Symbol or Formula: Not applicable [Physical parameter]

Units Used for Analytical Results: Degrees Celsius [°C].

Standard Method(s) of Analysis: Thermometry [A] or Thermistor [as in DO probe] [A/B], with measurement in the field (usually in association with the DO measurement: q.v.).

Occurrence/Origin: Generally climatologically influenced (in the absence of thermal discharges).

Health/Sanitary Significance: None.

Background Information: The natural variation in temperature found in Fresh surface waters is of the order of 25°C - from freezing point to a summer maximum of around 25°C in occasional years. Thermal pollution would, of course, alter the position, possibly vary significantly. The effect of temperature, especially temperature changes, on living organisms can be critical, and the subject is a vast and complex one. Where biochemical reactions are concerned, as in the uptake of oxygen by bacteria, a rise of 10°C in temperature leads to an approximate doubling of the reaction rate. Conversely, such reactions are retarded by cooling. Hence the recommendation often cools water to 4°C in the interval between sampling and analysis. Another critical factor is that some essential constituents of water either change their form (as in the ionization of ammonia) or alter their concentration (as with dissolved oxygen) when the temperature changes. However, elevated temperatures and, more importantly, steep temperature gradients can have directly harmful effects on fish (WHO, 2001).

2.4 Groundwater and public health

There are numerous endemic and epidemic diseases resulted from inadequate water supply. One of the reasons for increased deaths and preventable diseases in developed and developing countries is the rise of waterborne diseases. To provide health gains in people, enhancing water quality control plans, removing wastes, and enhancing personal hygiene should be done (Oliver Schmoll, 2006).

2.4.1 Effects of Ground Water Pollution

Groundwater is more preferred compared to surface water. This is due to its decreased cost, the ability of groundwater to be treated easily. Since it is inexpensive, hence it is used as a source of drinking water. Groundwater is used in the industry as it can give environmental benefits by recharging streams and rivers. When developing protection plans and strategies, implementing such measures should be considered, and the charge of not protecting Groundwater, for balanced decisions to be made (Oliver Schmoll, 2006).

Sudan relies on groundwater aquifers to supply water, both for human consumption and irrigation. It is determined that the groundwater potentialities of the basins are high. Furthermore, large amounts of Groundwater can be obtained for future growth in irrigation and domestic supply (Omer, 2002).

In contrast, it was found that stomach cancer is highest in areas where the groundwater concentration of NO_3^- is lowest and vice versa. Therefore, there is still no concrete evidence to associate NO_3^- contamination with stomach and gastrointestinal cancer. However, a study indicated that even at exposure levels of 111 mg/l, there were no adverse conditions in infants except for methemoglobinemia. Therefore, nitrate alone may not be the only cause of elevated regional gastric cancer mortality rates. These may result from several other factors, such as high pesticide levels, coliform bacteria, and other groundwater contaminants. Nitrate, nitrite, and many nitro have a carcinogenic effect on urothelial cells (Abdel-Magid, Yahia, & Abdellah, 2012). Furthermore, (Abdel-Magid, Yahia, & Abdellah, 2012) stated that there is a positive relationship reported between the NO_3^- level in drinking water and bladder cancer and ovarian cancer. Still, an inverse association between uterine cancer and rectal cancer is observed. There is a direct association between the dietary intake of NO_3^- and the incidence of urothelial cancers (renal pelvis, ureter, urinary bladder, and urethra), but no association for prostate cancer, renal tumors, or penile tumors have been found. However, epidemiological studies examining associations between cancer and nitrates in drinking water are inconclusive and are only circumstantial. It has been described that standards and guidelines for major cations and anions in drinking water were set based on taste considerations rather than the impact on human health. Thus, higher values can be tolerated. However, excessive concentrations of Na^+ in drinking groundwater can be a health risk factor for those on a low-sodium diet. Bacterial contamination of drinking water, directly and indirectly, affects human health and the environment. In general, there is an association between the microbial growth in drinking water with rainfall and water temperature more significant than 150C, which is much lower than the average temperature. On the other hand, (Abdel-Magid, Yahia, & Abdellah, 2012) attributed the absence of microbial contamination from Groundwater to complete human and animal contact protection.

The continuous changes in groundwater chemical constituents' levels necessitate a frequent examination of groundwater quality before being used.

2.4.2 Effects of drinking water on human health in Sudan

This study is conducted to investigate the following objectives:

1. To assess the groundwater quality correlated with health problems noticed recently in the study area.

2. To compare the results with the local, regional, and international standards and guidelines.

Groundwater quality criteria are explained by checking the components' concentrations and then determining whether they are high or not. Moreover, when exceeded, this will prohibit or impair water use (New Jersey Administrative Code, 2020).

Moreover, acording to (New Jersey Administrative Code, 2020), Hazardous pollutant means:

Any pollutant that causes toxicity.

Any substance considered as a pesticide under the Federal Insecticide, Fungicide, and Rodenticide.

Any substance in which its use or manufacturing is prohibited under the Federal Toxic Substances Control Act.

Any substance that is considered a carcinogen by the International Agency for Research on Cancer.

A pollutant is defined as any dredged spoil, solid waste, incinerator residue, sewage, garbage, refuse, oil, grease, sewage sludge, munitions, chemical wastes, biological materials, radioactive substance, thermal waste, wrecked or discarded equipment, rock, sand, cellar dirt, and industrial, municipal or agricultural or other residue discharged into the waters of the State (Code, 2016).

Pollutant includes both hazardous and non-hazardous pollutants (New Jersey Administrative Code, 2020).

Industrial, municipal or agricultural, or other residues specifically include, without limitation, constituents that are not considered wastes (that is, process chemicals) before discharge but are discharged and/or do degrade natural or existing groundwater quality.

On the other hand, the excessive F^- administration on human is dental fluorosis and skeletal fluorosis. It was confirmed that skeletal fluorosis could result from prolonged consumption of well water with more than four mg/l F⁻ in the drinking water. Furthermore, it has been reported that the kidneys are probably the most crucial organ in the human body during low-dose long-term exposure to F⁻ (Abdel-Magid, Yahia, & Abdellah, 2012). Healthy kidneys excrete 50 to 60% of the ingested dose. Kidney malfunction can impede this excretion, causing an increased F⁻ into bones and eventually causes skeletal fluorosis. Individuals with kidney disease have a decreased ability to excrete F⁻ in the urine. They are at risk of developing fluorosis even generally recommended in the past; it has offered insurance against drought, a vital factor for human survival. Thought once to be an inexhaustible resource for human needs, a belief reflected in many myths, Groundwater's availability is now critical in many parts of the world. It has become more challenging to supply human needs with sufficient quantities of safe and clean water of acceptable quality. GW is not only crucial in supporting human welfare; it is also the basis of life for diverse organisms existing below the earth's surface. The complex relationships between the water and the subsurface organisms generate dynamic ecological systems, the relatively young GW ecology field (DANIELOPOL, GRIEBLER, GUNATILAKA, & GUNATILAKA, 2003).

2.5 Groundwater as A Source of Drinking water

Groundwater has higher stability and microbial quality than surface waters. Groundwaters differ from surface water in the sense of treatments. Groundwater does not need treatment to be potable. On the other hand, surface waters require treatment.

Therefore, the quality of Groundwater must be protected if public health is not to be compromised.

Two factors cause Groundwater to have diversity in use. These factors are the land and ability to access sources of water. According to (Oliver Schmoll, 2006), 96 percent of domestic water comes from Groundwater in rural areas of the USA. In the United Kingdom, there is a variation in the usage of Groundwater. Generally, the national average for groundwater usage is 28 percent, whereas England's southern counties rely on Groundwater than the northern counties.

The countries that gain a high amount of municipal water supply Groundwater are Mexico City, Mexico, Lima, Peru, Buenos Aires, Argentina, and Santiago de Chile, and Chile. Regarding India, China, Bangladesh, Thailand, Indonesia, and Viet Nam, groundwater is considered more than 50 percent, representing potable supplies (Oliver Schmoll, 2006).

In Africa and Asia, most of Africa and Asia's largest cities prefer to use surface water. However, Groundwater is more found and used by people in rural areas and low-income peri-urban communities (Oliver Schmoll, 2006).

As a result, these populations are most vulnerable to waterborne disease. Approximately as much as 80 percent of the drinking-water used by these communities is abstracted from groundwater sources (Oliver Schmoll, 2006).

Groundwater, compared to surface water, has numerous advantages. It can be located in many areas and can be developed at a comparatively low cost.

To develop groundwater, minimal capital costs are required compared to surface water.

Disadvantages of surface water reservoirs include the inability of numerous functions to be operated for the optimum water supply benefit, such as water supply, flood control, irrigation, hydroelectric power, and recreation.

Layers of soil and sediment protect aquifers, which filter rainwater as it percolates through them, thus removing particles, pathogenic microorganisms, and many chemical constituents. Therefore it is generally assumed to be a relatively safe drinking-water source. However, Groundwater has been termed the 'hidden sea' – sea because of the large amount and hidden because it is not visible. This highlights a key issue in using aquifers as a drinking-water source, showing that particular attention is needed to ascertain whether Groundwater's general assumption is safe to drink valid in individual settings. As discussed below, understanding the source-pathway-receptor relationship in any particular location is critical to determine whether pollution will occur.

While there is a large volume of Groundwater in this 'hidden sea,' its replenishment occurs slowly – at rates varying between locations. Over-exploitation, therefore, readily occurs, bringing with it additional quality concern Groundwater and public health (Oliver Schmoll, 2006).

2.6 Disease derived from groundwater use

2.6.1 Chemical hazards

Two chemicals possess hazardous effects on Groundwater. These chemicals are fluoride and arsenic. First of all, fluoride interferes with bone and teeth development. Some diseases are consequences of having excess levels of fluoride. For example, skeletal fluorosis is a disease that leads to the inability of bones to function correctly. However, when water-containing fluoride is low in the body, this has also been connected with dental caries.

In addition to the above, arsenic is the largest recorded poisoning in history (Schmoll, Howard, Chilton, & Chorus, 2006).

2.6.2 Infectious disease transmission

Various infectious diseases are correlated with Groundwater. Those contagious diseases are produced from organisms like viruses and bacteria.

Moreover, such diseases play a significant role in elevating morbidity and mortality worldwide. Furthermore, most of those diseases are caused by common water-related pathogens such as typhoid and cholera. However, it is discovered that there are new pathogens and strains of established pathogens are being recognized. Regarding the factors that aid in making such pathogens emerge in water, agriculture and wastewater are among them. Demographic, behavioral changes, and socio-economic factors also lead to the emergence and re-emergence of pathogens in water (WHO, Emerging Issues in Water and Infectious Disease, 2003). The transmission of infectious diseases is connected to water and pathogen interactions.

They reveal a relationship depending on the pathogens' transmission characteristics and how water can help.

Regarding the classification of water-related infectious diseases, each disease was as follows: waterborne, water-based, water-related, water-washed, and water-dispersed. Firstly, waterborne diseases, such as typhoid and cholera, are mainly described by having microorganisms enter water sources through fecal contamination and cause infections in humans through contaminated water. Secondly, water-based diseases indicate diseases in which worms live in water sources then infect the human body, such as schistosomiasis. The third category is water-related diseases, such as malaria and trypanosomiasis. Microorganisms in this category breed vectors inside the water to complete the transmission cycle (WHO, Emerging Issues in Water and Infectious Disease, 2003).

3. Material and Methods

Study Design that utilized a cross-sectional study based on questionnaires and experimental work.

3.1 Materials

the materials which were used in the methodology consisted of:

a) questionnaires

b) 1-liter polyethylene plastic bottles

c) Spectrophotometer (Model 1100)

d) chromatograph 2000i

e) conductivity meter

f) pH meter

g) nitric acid

h) sulfuric acid solution

i) standard silver nitrate solution

j) potassium chromate solution

k) samples of water from various locations

l) red zirconium

3.2 Methodology

3.2.1 Data of the questionnaire

A 100-item paper questionnaire (see appendix) will be given to citizens from the targeted areas. The questionnaire will be based on two domains. The first domain is meant to obtain sociodemographic information. The second domain will be assessing an individual's knowledge about chemical water pollution. The questionnaire will be translated into Arabic and translated back to English to validate the Arabic version.

The questionnaire will be filled and returned to the person handing out the questionnaire within a specific time. The questionnaire will be formatted to adhere to the local culture's standard of phrasing the questionnaire and the participants' information.

Data will be analyzed using SPSS software, and graphs like scatter bar charts and pie charts will be used for this study.

1. Study Area

The study was conducted in Khartoum state (the capital of Sudan).

The questionnaires were distributed in:

1- Khartoum city

2- Omdurman city

3- Bahri city

Inclusion criteria:

1. Citizens

2. Teenagers to Adults; age from 15-60

3. Male or female, and

4. Agreement to participate in filling the questionnaire.

Sample Size Estimation:

A sample size of 100 participants was used

2. Experimental work

Wells water samples were collected from 23 locations (Figure 1). The water samples were collected in clean 1-liter polyethylene plastic bottles and stored in a cooler for 24 hours. Two samples location, two water samples were collected; one was acidified with nitric acid for anions determination. Electrical conductivity and pH were determined in the field.

Analysis for the cations Na^+, K^+, Ca^{2+}, Mg^{2+} was carried out using a Perkin Elmer Atomic Absorption Spectrophotometer (Model 1100).

The anions SO_4^{2-}, Cl^-, HCO_3^- and CO_3^{2-} were analyzed using a Dionex ion chromatograph 2000i.

3. Study Area and Duration

Twenty-three samples of drinking water were collected from 2012 to 2017 from different Khartoum, Omdurman, and Bahri areas.

This study concentrated on conducting a chemical analysis on samples of water taken from 23 wells in some areas near Khartoum's industrial areas, e.g., Hay Al Nuzha, Jabra, Al-Shagara, Al-Azouzab, Al-Lamb, Al-Kalakla, and Yathrib. In contrast, other samples were taken from Omdurman, e.g., Al-Shati and Bant, etc. The rest samples were taken from the city of Bahri, e.g., ElhagYousif, and El-Shaglah.

41

Table 5: Location of Study Area

No	Location	Long (E)	Lat (N)
1	Gabrah	32.514728	15.523675
2	ELShagara	32.500498	15.538726
3	ALKalaklah	32.480098	15.471675
4	Lamab	32.492500	15.540833
5	Abu Seid	32.483333	15.540833
6	ELHag Yousif	32.639049	15.641917
7	Elthora	32.492680	15.721746
8	Hay Elnozha	32.532462	15.552653
9	Hay Elshaty	32.505345	15.682158
10	Banat	32.482144	15.622884
11	Alhag Yosif Street 1	32.630609	15.666466
12	ElAzozab	32.489179	15.514846
13	Omdurman Hara 52	32.450138	15.734273
14	Gabra B9	32.525864	15.538647
15	OmbadaAmreya	32.396218	15.632796
16	ElDobaseen	32.490401	15.510220
17	Abu Seid B26	32.450923	15.593345
18	ElShaglah	32.666000	15.533000
19	ElDoma	32.493523	15.658749
20	ElBankElAgary	32.420752	15.596917
21	Elthora Hara 4	32.495457	15.679356
22	El Arda	32.472837	15.638887
23	Elthora El Eskan 52	32.456895	15.736481

This study has been done in some areas near industrial areas in Khartoum, as shown in the table (5).

Figure 1: Location of Study Area

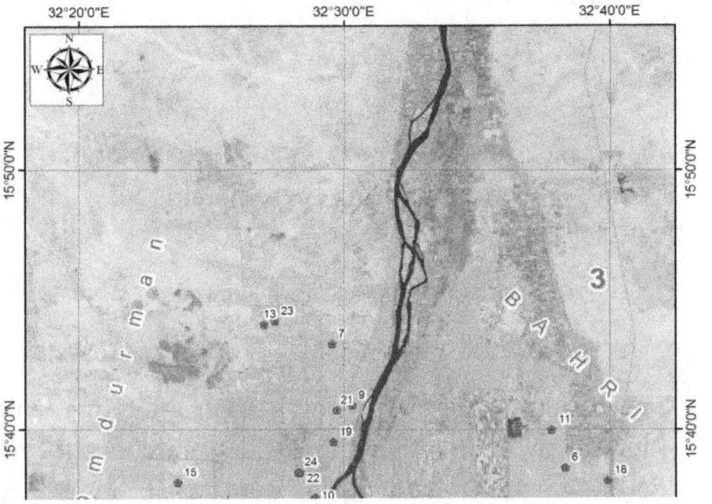

4. Chemical tests

Some chemical characteristics were carried out for drinking water samples. These analysis ionic components included:

- Anions: such as Cl^-, NO_3^- and SO_4^{2-}

- Cations: such as Na^+, Ca^{++} and Mg^{++}

1\hardness (total)

Dissolved minerals cause hardness in water, primarily divalent cations; calcium & Magnesium ions usually are the only ions that present an insignificant amount; therefore, hardness is generally considered a measure of the Calcium & Magnesium content of water.

The sample's pH can be adjusted to 10 with ammonium chloride solution or the addition of Eriochrome black T indicator followed by titration Vs. (EDTA disodium salt).

Result recorded as mg/l total hardness calculated as calcium carbonate.

2\ Calcium

Calcium hardness is determined after removing magnesium interference by adjusting the sample's pH to 12 with sodium hydroxide, then the amount of calcium is calculated.

The result was reported as mg/l calcium.

3\ Magnesium

Magnesium is determined by using the mathematical method by subtracting the calcium hardness from total hardness the remaining amount contributed to the magnesium the result reported as mg/l magnesium.

The chemical equation to treat hardness using titration with Aluminum sulfate Magnesium Sulphate and Calcium Sulphate will occur and easily can be removed as in the following equations :

1. $6Ca^{+2} + 2Al_2(SO4)_3 ---- 6Ca\,SO_4 + 4Al^{+3}$

2. $6Mg^{+2} + 2Al_2(SO4)_3 ---- 6Mg\,SO_4 + 4Al^{+3}$

4\ Total alkalinity

The alkalinity of water is water's capacity to neutralize acids. It is expressed as phenolphthalein & total; both types are determined with direct titration Vs. Standard sulfuric acid solution.

The result was reported in mg/l calcium carbonate.

5\ Chloride

Sample titrated Vs. Standard silver nitrate solution in the presence of potassium chromate as an indicator.

The result was reported in mg/l chloride.

6\ Nitrate

cadmium metal reduces nitrates present in the sample to nitrite. In acidic medium react with sulfanilic acid to form an intermediate diazonium salt, this salt couples with gentistic acid form an amber-colored product the intensity of color directly proportional to the amount of nitrate wavelength (λ) 500 nm.

Instrument: (HACH) 2000 DR spectrophotometer.

Result reported in mg/l nitrate

Ref\ (G.H. Jeffery, J. Bassett, J. Mendham, and R.C. Denney, Quantitative inorganic chemistry, Sixth edition, 2000)

7\ Nitrite

In the sample, nitrite reacts with sulfanilic acid to form an intermediate Diazonium salt couple with chronotropic acid to form pink colored the intensity of color in proportional to the amount of nitrite in the sample.

Wavelength (λ) 507 nm

Instrument: (HACH) 2000 DR spectrophotometer.

Result reported in mg/l nitrite

(EPA) approve.

8\ Ammonia

Nessler method

Nessler reagent reacts under strongly alkaline conditions with the ammonia present in the sample to produce yellow-colored species. The intensity of the color is proportional to the amount of ammonia in the sample.

Wavelength (λ) 425 nm.

Instrument: (HACH) 2000 DR spectrophotometer.

Result reported in mg/l ammonia

(EPA) approved.

9\ Hydrogen sulfide & sulfides

Hydrogen sulfide & acid-soluble sulfides react with NN-dimethyl p-phenylenediamine sulfate) to form methylene blue, the color's intensity is proportional to the amount of sulfide in the sample.

Wavelength (λ) 665 nm.

Instrument: (HACH) 2000 DR spectrophotometer.

Result reported in mg/l hydrogen sulfide

10\ Sulfate

Sulfate in the sample reacts with barium chloride to form a precipitate of barium sulfate. The amount of turbidity formed is directly in proportion to the amount of sulfate.

Wavelength (λ) 450 nm.

Instrument: (HACH) 2000 DR spectrophotometer.

Result reported in mg/l sulfate

Ref\ (Lenore S. Clesceri, Standard Method for The Examination of Water & Waste Water, 1998)

11\ Fluoride

SPANDS method

This method involves the reaction of fluoride with red zirconium–dye solution. The fluoride reacts with part of the zirconium to form a colorless complex, thus bleaching the red color according to the concentration of fluoride in the sample. Wavelength (λ)580 nm.

Instrument: (HACH) 2000 DR spectrophotometer.

Result reported in mg/l fluoride

Ref\ (Lenore S. Clesceri, Standard Method for The Examination of Water & Waste Water, 1998)

12\ Iron ferrous

1,10 phenanthroline method

1,10 phenanthroline react with ferrous to form an orange color the

the intensity of color is proportional to the amount of ferrous in the sample

Wavelength (λ) 510 nm. Result reported in mg/l ferrous iron

Instrument: (HACH) 2000 DR spectrophotometer

Ref\ (Lenore S. Clesceri, Standard Method for The Examination of Water & Waste Water, 1998)

13\ Iron total

Forever method

forever reagent converts all soluble & most insoluble forms of iron in the sample to soluble ferrous iron, which reacts with 1,10 phenanthroline in the reagent to form an orange color, the intensity of color is proportional to the amount of iron in the sample.

Result reported in mg/l total iron, Wavelength (λ) 510 nm.

Instrument: (HACH) 2000 DR spectrophotometer

Ref\ (Lenore S. Clesceri, Standard Method for The Examination of Water & Waste Water, 1998)

14\ Total dissolved solids (TDS)

Direct reading by using conductivity (TDS) meter

The result was reported in mg/l (TDS).

15\ Free chlorine

Chlorine in the sample reacts with DPD free chlorine powder pillow.

The reagent to forman pink color the intensity of color in proportional to the amount of free chlorine in the sample result reported in mg/L.

Wavelength (λ) 530 nm [6].

---

[6] Ref: Lenore S. Clesceri, Standard Method for The Examination of Water & Waste Water, 1998.

4. Result

Statistical Methods: A comparative statistical analytical method using the SPSS statistical program based on the Chi-square test and Kruskal-Wallis test.

Table 6: Frequency distribution of participants according to Age:

Age	Frequency	Percentage	Valid Percentage
less than 20 Years	16	16.0	16.2
20-30 Years	60	60.0	60.6
31-40 Years	8	8.0	8.1
41-50 Years	11	11.0	11.1
51-60 Years	2	2.0	2.0
More than 60 years	2	2.0	2.0
Total	99	99.0	100.0
Not determined	1	1.0	
Total	100	100.0	

Table (6) showed that most (60%) of participants were 20-30 years old. On the other hand, only (2%) of them were 51-60 years old or more than 60 years. While (1%) of the participants did not determine their age.

Table 7: Frequency distribution of participants according to Gender:

Gender	Frequency	Percentage	Valid Percentage
Female	48	48.0	48.0
Male	52	52.0	52.0
Total	100	100.0	100.0

Table (7) showed that (52%) of participants were females, while (48%) of them were males.

Table 8: Frequency distribution of participants according to Residence Place:

Residence	Frequency	Percentage	Valid Percentage
Bahri	25	25.0	25.0
Omdurman	36	36.0	36.0
Khartoum	36	36.0	36.0
Other	3	3.0	3.0
Total	100	100.0	100.0

Table (8) showed that most (36%) of participants live in Omdurman or Khartoum. However, (25%) of them live in Bahri, while only (3%) of them are from other places.

Table 9: Frequency distribution of participants according to Level of Education:

Level of Educ.	Frequency	Percentage	Valid Percentage
High school	5	5.0	5.0
College or above	95	95.0	95.0
Total	100	100.0	100.0

Table (9) showed that only (5%) of participants had a high school level of education, whereas the majority (95%) of them had a college level of education.

Table 10: Frequency distribution of participants according to Marital Status:

Marital status	Frequency	Percentage	Valid Percentage
Married	19	19.0	19.4
Single	77	77.0	78.6
Widow/divorced	2	2.0	2.0

Total	98	98.0	100.0
Not determined	2	2.0	
Total	100	100.0	

Table (10) showed that most (77%) of participants were single, while (19%) were married, and only (2%) were Widow/divorced or did not determine.

Table (11) Frequency distribution of participants according to if they have ever heard about water pollution by chemicals with respect to their ages:

			Have you ever heard about water pollution by chemicals?		
			Yes	No	Not sure
Age	less than 20 Years	Count	13	1	2
		N %	81.20%	6.20%	12.50%
	20-30 Years	Count	57	1	2
		N %	95.00%	1.70%	3.30%
	31-40 Years	Count	7	0	1
		N %	87.50%	0.00%	12.50%
	41-50 Years	Count	8	1	2
		N %	72.70%	9.10%	18.20%
	51-60 Years	Count	2	0	0
		N %	100.00%	0.00%	0.00%
	More than 60 years	Count	2	0	0
		N %	100.00%	0.00%	0.00%
Independence	Chi-Square		6.993		
	Df		5		
	Asymp. Sig.		0.221		

Table (11) showed that most of the participants with different age categories heard about water pollution.

Table (12) Frequency distribution of participants according to their source of information about water pollution by chemicals with respect to their ages:

7			What was your source of information about water pollution by chemicals?			
			Water pollution preventer	Social media	Physician	Other
Age	less than 20 Years	Count	0	8	1	7
		N %	0.00%	50.00%	6.20%	43.80%
	20-30 Years	Count	8	34	8	10
		N %	13.30%	56.70%	13.30%	16.70%
	31-40 Years	Count	3	2	1	2
		N %	37.50%	25.00%	12.50%	25.00%
	41-50 Years	Count	2	7	2	0
		N %	18.20%	63.60%	18.20%	0.00%
	51-60 Years	Count	0	1	0	1
		N %	0.00%	50.00%	0.00%	50.00%
	More than 60 years	Count	1	0	1	0
		N %	50.00%	0.00%	50.00%	0.00%
Independence	Chi-Square		7.4			
	df		5			
	Asymp. Sig.		0.193			

Table (12) showed that most participants with different age categories used social media as their source of information about water pollution by chemicals. But, for those whose ages between 31-40 years, their sources were water pollution preventers.

Table (13) Frequency distribution of participants according to their understanding of water pollution with respect to their ages:

			Do you know anything about water pollution?			
			No knowledge	Low knowledge	Moderate knowledge	Well knowledge
Age	less than 20 Years	Count	2	4	7	3
		N %	12.50%	25.00%	43.80%	18.80%
	20-30 Years	Count	3	10	28	19
		N %	5.00%	16.70%	46.70%	31.70%
	31-40 Years	Count	0	3	3	2
		N %	0.00%	37.50%	37.50%	25.00%

	41-50	Count	0	2	5	4
	Years	N %	0.00%	18.20%	45.50%	36.40%
	51-60	Count	0	0	1	1
	Years	N %	0.00%	0.00%	50.00%	50.00%
	More than 60 years	Count	0	1	0	1
		N %	0.00%	50.00%	0.00%	50.00%
Independence	Chi-Square		3.486			
	Df		5			
	Asymp. Sig.		0.625			

Table (13) showed most participants with different age categories had moderate knowledge about water pollution. On the contrary, for those whose ages between 51-60 Years, half of them had moderate knowledge, and the other half have well knowledge.

Table (14) Frequency distribution of participants according to their drinking water source with respect to their ages:

			What is the source of your drinking water?			
			Nile river	Dug well	Drilled well	Not sure
Age	less than 20 Years	Count	9	0	1	6
		N %	56.20%	0.00%	6.20%	37.50%
	20-30 Years	Count	44	4	4	8
		N %	73.30%	6.70%	6.70%	13.30%
	31-40 Years	Count	4	2	0	2
		N %	50.00%	25.00%	0.00%	25.00%
	41-50 Years	Count	10	0	1	0
		N %	90.90%	0.00%	9.10%	0.00%
	51-60 Years	Count	2	0	0	0
		N %	100.00%	0.00%	0.00%	0.00%
	More than 60 years	Count	2	0	0	0
		N %	100.00%	0.00%	0.00%	0.00%
Independence	Chi-Square		8.004			
	Df		5			

	Asymp. Sig.	0.156

Table (14) showed most participants of different age categories; their source of drinking water was the Nile River.

Table (15) Frequency distribution of participants according to the Color of their Drinking Dater with respect to their ages:

			What is the color of your drinking water?					
			White	Silver	Green	Yellow	Colorless	Other
Age	less than 20 Years	Count	2	1	1	2	9	1
		N %	12.50 %	6.20 %	6.20 %	12.50 %	56.20%	6.20%
	20-30 Years	Count	10	1	1	7	33	8
		N %	16.70 %	1.70 %	1.70%	11.70 %	55.00%	13.30 %
	31-40 Years	Count	2	0	1	1	2	2
		N %	25.00 %	0.00 %	12.50 %	12.50 %	25.00%	25.00 %
	41-50 Years	Count	3	0	0	6	2	0
		N %	27.30 %	0.00 %	0.00 %	54.50 %	18.20%	0.00%
	51-60 Years	Count	0	0	0	0	2	0
		N %	0.00%	0.00 %	0.00 %	0.00%	100.00 %	0.00%
	More than 60 years	Count	1	0	0	0	1	0
		N %	50.00 %	0.00 %	0.00 %	0.00%	50.00%	0.00%
Independence	Chi-Square		7.517					
	Df		5					
	Asymp. Sig.		0.185					

Table (15) showed most participants with different age categories had colored drinking water. However,

54

regarding the participants whose ages between 51-60 years, their drinking water was colorless.

Table (16) Frequency distribution of participants according to if their drinking water has an Odor with respect to their ages:

			Does your drinking water have an odor?		
			Yes	No	Not sure
Age	less than 20 Years	Count	2	14	0
		N %	12.50%	87.50%	0.00%
	20-30 Years	Count	13	42	3
		N %	22.40%	72.40%	5.20%
	31-40 Years	Count	2	6	0
		N %	25.00%	75.00%	0.00%
	41-50 Years	Count	4	6	1
		N %	36.40%	54.50%	9.10%
	51-60 Years	Count	0	2	0
		N %	0.00%	100.00%	0.00%
	More than 60 years	Count	0	1	1
		N %	0.00%	50.00%	50.00%
Independence	Chi-Square		4.389		
	Df		5		
	Asymp. Sig.		0.495		

Table (16) showed most participants with different age categories had odorless drinking water. But, regarding the participants whose age was more than 60 years, half had odorless water, and the other half were not sure.

Table (17) Frequency distribution of participants according to if their drinking water has a taste with respect to their ages:

55

			Does your drinking water have a taste?		
			Yes	No	Not sure
Age	less than 20 Years	Count	3	12	1
		N %	18.80%	75.00%	6.20%
	20-30 Years	Count	16	40	4
		N %	26.70%	66.70%	6.70%
	31-40 Years	Count	5	3	0
		N %	62.50%	37.50%	0.00%
	41-50 Years	Count	3	6	2
		N %	27.30%	54.50%	18.20%
	51-60 Years	Count	0	2	0
		N %	0.00%	100.00%	0.00%
	More than 60 years	Count	0	2	0
		N %	0.00%	100.00%	0.00%
Independence	Chi-Square		6.266		
	Df		5		
	Asymp. Sig.		0.281		

Table (17) showed most participants with different age categories had tasteless drinking water. While the participants whose age was between 31-40 years, the majority of them had water with a taste.

Table (18) Frequency distribution of participants according to if they experience drinking water cut with respect to their ages:

			Do you experience drinking water cut?		
			Yes	No	Not sure
Age	less than 20 Years	Count	5	8	3
		N %	31.20%	50.00%	18.80%
	20-30 Years	Count	22	27	11
		N %	36.70%	45.00%	18.30%
	31-40 Years	Count	6	2	0
		N %	75.00%	25.00%	0.00%
	41-50 Years	Count	6	2	3
		N %	54.50%	18.20%	27.30%
	51-60 Years	Count	1	1	0
		N %	50.00%	50.00%	0.00%

	More than 60 years	Count	2	0	0
		N %	100.00%	0.00%	0.00%
Independence	Chi-Square		7.541		
	Df		5		
	Asymp. Sig.		0.183		

Table (18) showed the majority of participants with different age categories experienced drinking water cut.

Table (19) Frequency distribution of participants according if they boil water before using it with respect to their ages:

			Do you boil water before use	
			Yes	No
Age	less than 20 Years	Count	2	14
		N %	12.50%	87.50%
	20-30 Years	Count	12	48
		N %	20.00%	80.00%
	31-40 Years	Count	3	5
		N %	37.50%	62.50%
	41-50 Years	Count	4	7
		N %	36.40%	63.60%
	51-60 Years	Count	0	2
		N %	0.00%	100.00%
	More than 60 years	Count	0	2
		N %	0.00%	100.00%
Independence	Chi-Square		4.59	
	Df		5	
	Asymp. Sig.		0.468	

Table (19) revealed that most participants with different age categories did not boil water before using it.

Table (20) Frequency distribution of participants according to the method they use to purify water with respect to their ages:

			By which method do you purify water?			
			Filtration	Chemically	Both filtration and chemical	None
Age	less than 20 Years	Count	6	1	3	6
		N %	37.50%	6.20%	18.80%	37.50%
	20-30 Years	Count	22	1	6	30
		N %	37.30%	1.70%	10.20%	50.80%
	31-40 Years	Count	2	0	0	6
		N %	25.00%	0.00%	0.00%	75.00%
	41-50 Years	Count	6	1	0	4
		N %	54.50%	9.10%	0.00%	36.40%
	51-60 Years	Count	1	0	0	1
		N %	50.00%	0.00%	0.00%	50.00%
	More than 60 years	Count	1	0	0	0
		N %	100.00%	0.00%	0.00%	0.00%
Independence	Chi-Square	4.293				
	Df	5				
	Asymp. Sig.	0.508				

Table (20) showed some participants with different age categories used filtration methods to purify their drinking water. On the other hand, a few of them used chemical methods. Moreover, a small percentage of the participants used both filtration and chemical methods. The rest of the participants did not use any method.

Table (21) Frequency distribution of participants according to the use of water with respect to their ages:

			What do you use water for?		
			Domestic	Domestic and agricultural uses	Other
Age	less than 20 Years	Count	10	4	2
		N %	62.50%	25.00%	12.50%
	20-30 Years	Count	37	21	2
		N %	61.70%	35.00%	3.30%
	31-40 Years	Count	3	5	0
		N %	37.50%	62.50%	0.00%
	41-50 Years	Count	4	6	1
		N %	36.40%	54.50%	9.10%
	51-60 Years	Count	2	0	0
		N %	100.00%	0.00%	0.00%
	More than 60 years	Count	2	0	0
		N %	100.00%	0.00%	0.00%
Independence	Chi-Square		6.304		
	Df		5		
	Asymp. Sig.		0.278		

Table (21) showed some participants with different age categories utilized water for domestic purposes. Furthermore, some participants used water for domestic and agricultural purposes. Moreover, the age categories (from less than 20 years to 50 years) utilized water for other uses. However, the age categories (51-60 years and more than 60 years) did not use water for other purposes.

Table (22) Frequency distribution of participants according to whether people or one of their relatives suffer from a waterborne disease with respect to their ages:

			Do you or one of your relatives suffer from a waterborne disease?		
			Yes	No	Not sure
Age	less than 20 Years	Count	1	11	4
		N %	6.20%	68.80%	25.00%
	20-30 Years	Count	21	32	7
		N %	35.00%	53.30%	11.70%
	31-40 Years	Count	3	2	3
		N %	37.50%	25.00%	37.50%
	41-50 Years	Count	4	5	2
		N %	36.40%	45.50%	18.20%
	51-60 Years	Count	0	2	0
		N %	0.00%	100.00%	0.00%
	More than 60 years	Count	0	1	1
		N %	0.00%	50.00%	50.00%
Independence	Chi-Square		7.169		
	Df		5		
	Asymp. Sig.		0.208		

Table (22) showed some participants with different age categories, and one of their relatives suffered from a waterborne disease. But, some participants did not suffer from any waterborne diseases. However, the minority was not sure. Regarding the participants whose age category is (more than 60 years) in addition to their relatives, they revealed that they did not suffer from any waterborne disease.

Table (23) Frequency distribution of participants according to if they have ever heard about water pollution by chemicals with respect to their residence:

		Have you ever heard about water pollution by chemicals?		
		Yes	No	Not

						sure
Residence	Bahri	Count	23	0	2	
		N %	92.00%	0.00%	8.00%	
	Omdurman	Count	32	1	3	
		N %	88.90%	2.80%	8.30%	
	Khartoum	Count	33	1	2	
		N %	91.70%	2.80%	5.60%	
	Other	Count	2	1	0	
		N %	66.70%	33.30%	0.00%	
Independence	Chi-Square		1.704			
	Df		3			
	Asymp. Sig.		0.636			

Table (23) indicated that the majority of participants have heard about water pollution caused by chemicals. A few of them either revealed that they did not hear about this and some of them were not sure.

Table (24) Frequency distribution of participants according to their source of information about water pollution by chemicals with respect to their residence:

			What was your source of information about water pollution by chemicals?			
			Water pollution preventer	Social media	Physician	Other
Residence	Bahri	Count	3	14	5	3
		N %	12.00%	56.00%	20.00%	12.00%
	Omdurman	Count	7	16	3	10
		N %	19.40%	44.40%	8.30%	27.80%
	Khartoum	Count	3	21	5	7
		N %	8.30%	58.30%	13.90%	19.40%
	Other	Count	1	1	0	1

			N %	33.30%	33.30%	0.00%	33.30%
Independenc e	Chi-Square			0.264			
	Df			3			
	Asymp. Sig.			0.967			

Table (24) showed most participants considered social media as their source of information about water pollution by chemicals. On the contrary, a few of them chose water pollution preventer as their source. There was a small percentage revealed that physicians were their source of information. Finally, the rest of the participants used other sources.

Table (25) Frequency distribution of participants according to believe their understanding with respect to their residence:

			How well do you believe your understanding?			
			No knowledge	Low knowledge	Moderate knowledge	Well knowledge
Residence	Bahri	Count	0	8	11	6
		N %	0.00%	32.00%	44.00%	24.00%
	Omdurman	Count	2	3	19	12
		N %	5.60%	8.30%	52.80%	33.30%
	Khartoum	Count	2	9	13	12
		N %	5.60%	25.00%	36.10%	33.30%
	Other	Count	1	0	2	0
		N %	33.30%	0.00%	66.70%	0.00%
Independence	Chi-Square		2.909			
	Df		3			
	Asymp. Sig.		0.406			

Table (25) revealed that most participants, concerning their residence, had moderate water pollution knowledge.

Table (26) the frequency distribution of participants according to their drinking water source with respect to their residence:

			What is your drinking water source?			
			Nile river	Dug well	Drilled well	Not sure
Residence	Bahri	Count	19	0	0	6
		N %	76.00%	0.00%	0.00%	24.00%
	Omdurman	Count	29	2	2	3
		N %	80.60%	5.60%	5.60%	8.30%
	Khartoum	Count	22	3	4	7
		N %	61.10%	8.30%	11.10%	19.40%
	Other	Count	2	1	0	0
		N %	66.70%	33.30%	0.00%	0.00%
Independence	Chi-Square		3.333			
	Df		3			
	Asymp. Sig.		0.343			

Table (26) illustrated that the highest percentages of the participants concerning their residence chose the Nile River as their main drinking water source.

Table (27) Frequency distribution of participants according to the color of their water with respect to their residence:

			What is the color of your water?					
			White	Silver	Green	Yellow	Colorless	Other
Residence	Bahri	Count	2	1	0	6	12	4
		N %	8.00	4.0	0.00	24.0	48.00	16.0

			%	0%	%	0%	%	0%
	Omdurman	Count	9	0	1	3	18	5
		N %	25.00%	0.00%	2.80%	8.30%	50.00%	13.90%
	Khartoum	Count	7	1	1	7	18	2
		N %	19.40%	2.80%	2.80%	19.40%	50.00%	5.60%
	Other	Count	0	0	1	0	1	1
		N %	0.00%	0.00%	33.30%	0.00%	33.30%	33.30%
Independence	Chi-Square	1.894						
	Df	3						
	Asymp. Sig.	0.595						

Table (27) showed that most of the participants, for their residence, indicated that their drinking water was colorless. The participants who lived in areas other than Bahri, Omdurman, and Khartoum revealed different colors.

Table (28) the frequency distribution of participants according to if their drinking water has an odor with respect to their residence:

			Does your drinking water have an Odor?		
			Yes	No	Not sure
Residence	Bahri	Count	6	15	3
		N %	25.00%	62.50%	12.50%
	Omdurman	Count	5	29	2
		N %	13.90%	80.60%	5.60%
	Khartoum	Count	8	27	0
		N %	22.90%	77.10%	0.00%
	Other	Count	2	1	0
		N %	66.70%	33.30%	0.00%
Independence	Chi-Square		4.898		
	df		3		

	Asymp. Sig.	0.179	

Table (28) showed most participants, based on their residence, revealed that their drinking water had no odor. However, the participants who lived in other areas determined that their drinking water had an odor.

Table (29) Frequency distribution of participants according to if their drinking water has a taste with respect to their residence:

			Does your drinking water have a taste?		
			Yes	No	Not sure
Residence	Bahri	Count	10	14	1
		N %	40.00%	56.00%	4.00%
	Omdurman	Count	8	27	1
		N %	22.20%	75.00%	2.80%
	Khartoum	Count	9	22	5
		N %	25.00%	61.10%	13.90%
	Other	Count	1	2	0
		N %	33.30%	66.70%	0.00%
Independence	Chi-Square		2.997		
	Df		3		
	Asymp. Sig.		0.392		

Table (29) showed that most participants, based on their residence, used tasteless drinking water. However, some of the participants had drinking water with a taste. Whereas a small percentage of the participants indicated that some participants were not sure about the taste.

Table (30) Frequency distributions of participants according to if they have drinking water cuts with respect to their residence:

			Do you have water cut?		
			Yes	No	Not sure
Residence	Bahri	Count	11	9	5
		N %	44.00%	36.00%	20.00%
	Omdurman	Count	15	17	4
		N %	41.70%	47.20%	11.10%
	Khartoum	Count	14	13	9
		N %	38.90%	36.10%	25.00%
	Other	Count	2	1	0
		N %	66.70%	33.30%	0.00%
Independence	Chi-Square		1.7		
	Df		3		
	Asymp. Sig.		0.637		

Table (30) showed most of the participants based on their residence had drinking water cuts.

Table (31) Frequency distribution of participants according if they boil water before use with respect to their residence:

			Do you boil water used for?	
			Yes	No
Residence	Bahri	Count	8	17
		N %	32.00%	68.00%
	Omdurman	Count	9	27
		N %	25.00%	75.00%
	Khartoum	Count	5	31
		N %	13.90%	86.10%
	Other	Count	0	3
		N %	0.00%	100.00%
Independence	Chi-Square		3.833	
	Df			

			3			
	Asymp. Sig.		0.28			

Table (31) showed that most participants based on their residence did not boil their drinking water.

Table (32) Frequency distribution of participants according to the method they use to purify water with respect to their residence:

			By which method do you purify water?			
			Filtration	Chemically	Both filtration and chemical	None
Residence	Bahri	Count	9	1	3	12
		N %	36.00%	4.00%	12.00%	48.00%
	Omdurman	Count	13	1	4	16
		N %	38.20%	2.90%	11.80%	47.10%
	Khartoum	Count	16	1	2	17
		N %	44.40%	2.80%	5.60%	47.20%
	Other	Count	0	0	1	2
		N %	0.00%	0.00%	33.30%	66.70%
Independence	Chi-Square		1.322			
	Df		3			
	Asymp. Sig.		0.724			

Table (32) showed that most participants, based on their residence, did not use any method to purify their water. Simultaneously, the second-highest percentages refer to the participants who used filtration methods to purify water. The third highest percentages indicated that some participants used both filtration and chemical methods. The minority of the participants used chemical methods to purify their water.

Table (33) Frequency distribution of participants according to uses of water with respect to their residence:

			What does the water use for?		
			Domestic	Domestic and agricultural uses	Other
Residence	Bahri	Count	13	11	1
		N %	52.00%	44.00%	4.00%
	Omdurman	Count	20	12	4
		N %	55.60%	33.30%	11.10%
	Khartoum	Count	24	12	0
		N %	66.70%	33.30%	0.00%
	Other	Count	1	2	0
		N %	33.30%	66.70%	0.00%
Independence	Chi-Square		2.665		
	Df		3		
	Asymp. Sig.		0.446		

Table (33) showed that most participants, based on their residence, used water for domestic purposes. Some of the participants used water for Domestic and agricultural uses. The minor percentages referred to the participants who used water for other purposes.

Table (34) Frequency distribution of participants according to whether people or one of their relatives suffer from a waterborne disease with respect to their residence:

			Do you or one of your relatives suffer from a waterborne disease?		
			Yes	No	Not sure
Residence	Bahri	Count	5	14	6
		N %	20.00%	56.00%	24.00%
	Omdurman	Count	7	24	5
		N %	19.40%	66.70%	13.90%
	Khartoum	Count	15	15	6
		N %	41.70%	41.70%	16.70%
	Other	Count	2	0	1
		N %	66.70%	0.00%	33.30%
Independence	Chi-Square		3.647		
	Df		3		
	Asymp. Sig.		0.302		

Table (34) showed that most of the participants and their relatives, based on their residence, did not suffer from any waterborne disease. But, some of the participants suffered from such types of diseases.

Table (35): Frequency distribution of study areas according to their water's color and its PH. Alkalinity:

			Color	PH. Alkalinity
			Nil	Nil
Area	Khartoum	Count	8	8
		% within Area	100.0%	100.0%
	Omdurman	Count	15	15
		% within Area	100.0%	100.0%
	Bahri	Count	3	3
		% within Area	100.0%	100.0%
Total		Count	26	26
		% within Area	100.0%	100.0%

Table (35) showed that drinking water in all areas had colored with PH. Alkalinity.

Table (36): The frequency distribution of study areas according to their water's odor and Chi-square test for independence:

			Odor		
			Nil	+ve	Total
Area	Khartoum	Count	7	1	8
		% within Area	87.5%	12.5%	100.0%
	Omdurman	Count	13	2	15
		% within Area	86.7%	13.3%	100.0%
	Bahri	Count	3	0	3
		% within Area	100.0%	.0%	100.0%
Total		Count	23	3	26
		% within Area	88.5%	11.5%	100.0%

Chi-Square Tests

	Value	Df	Asymp. Sig. (2-sided)
Pearson Chi-Square	0.446	2	0.800
Likelihood Ratio	0.788	2	0.674

Table (36) showed that the highest percentages of participants represented using drinking water with no odor in Khartoum, Omdurman, and Bahri.

Table (37): the frequency distribution of study areas according to their water's Appearance and Chi-square test for independence:

			Appearance		
			Clear	Turbid	Total
Area	Khartoum	Count	5	3	8
		% within Area	62.5%	37.5%	100.0%
	Omdurman	Count	14	1	15
		% within Area	93.3%	6.7%	100.0%
	Bahri	Count	3	0	3
		% within Area	100.0%	.0%	100.0%

Total	Count	22	4	26
	% within Area	84.6%	15.4%	100.0%

Chi-Square Tests

	Value	Df	Asymp. Sig. (2-sided)
Pearson Chi-Square	4.427	2	0.109
Likelihood Ratio	4.392	2	0.111

Table (37) showed that drinking water in Khartoum, Omdurman, and Bahri had a clear appearance, which is considered the highest percentages. But, small percentages revealed that there was a turbid appearance.

Table (38): the means, standard deviation, and standard error of parameters in study areas and Kruskall-Wallis test:

Parameter	Area	Mean	Std. Deviation	Std. Error	Kruskall-Wallis test	Df	Asymp. Sig.
Urbidity	Khartoum	8.1712	13.01330	4.60089			
	Omdurman	17.5427	40.05637	10.34251	0.203	2	0.904
	Bahri	2.3533	1.13161	.65333			
pH	Khartoum	7.47500	.353553	.125000			
	Omdurman	7.68000	.245531	.063396	2.121	2	0.346
	Bahri	7.56667	.305505	.176383			
Temperature	Khartoum	26.738	3.4159	1.2077			
	Omdurman	29.127	3.5684	.9214	3.376	2	0.185
	Bahri	24.467	3.3382	1.9273			
E_Conductivity	Khartoum	453.462	240.497930	85.02886			
	Omdurman	767.009	593.398448	153.2148	2.644	2	0.267
	Bahri	978.600	845.077677	487.9058			
T.D.S	Khartoum	260.275	128.700669	45.50256			
	Omdurman	425.809	324.268308	83.72572	2.257	2	0.324
	Bahri	489.300	422.538838	24.39529			
T.Alkalinity	Khartoum	162.500	67.551039	23.88290	1.032	2	0.597
	Omdurman	191.400	71.097318	18.35725			

	Bahri	171.000	9.643651	5.567764			
T.Hardness	Khartoum	138.250	49.551560	17.51912			
	Omdurman	184.667	75.983708	19.6189	2.633	2	0.268
	Bahri	307.067	314.356825	181.4940			
Phosphate	Khartoum	.2888	.15376	.05436			
	Omdurman	.3523	.44527	.11497	0.651	2	0.722
	Bahri	.2067	.04163	.02404			
Chloride	Khartoum	23.25	26.212	9.267			
	Omdurman	67.22	67.836	17.515	4.909	2	0.086
	Bahri	115.33	151.596	87.524			
Fluoride	Khartoum	.2662	.21071	.07450			
	Omdurman	.4633	.23690	.06117	6.47	2	0.039
	Bahri	.4700	.13077	.07550			
Sulfate	Khartoum	12.2500	13.53039	4.78371			
	Omdurman	52.6333	58.84317	15.19324	9.378	2	0.009
	Bahri	108.00	125.90473	72.69113			
Ammonia	Khartoum	.50000	.680882	.240728			
	Omdurman	.33427	.471514	.121744	1.328	2	0.515
	Bahri	.12365	.205574	.118688			
Nitrite	Khartoum	.06712	.106392	.037615			
	Omdurman	.13728	.309463	.079903	0.308	2	0.857
	Bahri	.02400	.013454	.007767			
Nitrate	Khartoum	6.52500	2.709639	.958002			
	Omdurman	4.90933	2.138232	.552089	2.819	2	0.244
	Bahri	11.7467	10.706285	6.181276			
Iron	Khartoum	.0462	.02504	.00885			
	Omdurman	.0912	.07909	.02042	1.446	2	0.485
	Bahri	.1967	.26502	.15301			
Calcium	Khartoum	31.3000	10.677078	3.774917			
	Omdurman	41.2800	24.588592	6.348747	2.002	2	0.368
	Bahri	48.9067	50.820174	29.34104			
Magnesium	Khartoum	14.3250	5.84190	2.06542			
	Omdurman	19.5460	8.12804	2.09865	3.965	2	0.138
	Bahri	44.3500	44.95750	25.95623			

Sodium	Khartoum	31.5912	25.058481	8.859511			
	Omdurman	70.7200	66.525610	17.17684	4.021	2	0.134
	Bahri	55.8300	36.787847	21.23947			
Potassium	Khartoum	5.7081	3.19384	1.12919			
	Omdurman	5.9973	2.81279	.72626	6.42	2	0.04
	Bahri	2.2067	.58432	.33736			
Manganes	Khartoum	.04925	.052169	.018445			
	Omdurman	.04288	.030056	.007760	0.078	2	0.962
	Bahri	.03833	.009815	.005667			

5. Discussion, Conclusion, and Recommendations

5.1. Discussion

1. Analysis results of the questionnaire

The Questionnaire results showed that about 72% live in Khartoum or Omdurman, while (25%) live in Bahri. Also, (3%) were from other places.

Tables (11-22) showed that most participants of all age categories have heard about water pollution by chemicals. The source of information about water pollution by chemicals is the social media for the 30-year category and less and 41-60 year category. However, the source of information for the 31-40 year category or more than 60-year category is water pollution preventer, which is not statistically different. Moreover, most of the participants have moderate knowledge about pollution by chemicals. The source of drinking water for most Nile River participants is colorless and does not have odor and taste, but it has water cuts. Most of the participants do not boil their water before use.

On the contrary, most of the participants aged less than 20 years or more than 40 years purify water by filtration. But, most of those aged 20-40 years do not purify water. Furthermore, most of the participants aged less than 30 years or more than 50 years use water for domestic purposes. In comparison, most of those aged 31-50 years use water for domestic and agricultural uses, but there were no significant statistical differences. Regarding waterborne diseases, most of the participants and one of their relatives suffer from a waterborne disease.

Table (23-34) showed that most of the participants living in multiple areas have heard about water pollution by chemicals. The source of information about water pollution by chemicals is social media. Additionally, most of the participants have moderate knowledge about pollution by chemicals. Furthermore, the source of their drinking water for most of them is the Nile River. According to participants from Bahri, Omdurman, and Khartoum, the Nile River is colorless and has no odor and taste. Whereas, in other residential areas, the Nile River has an odor, but statistically without significant differences.

Moreover, there is no water cut in Omdurman. Also, most participants do not boil water before use, and they do not purify it. Most of the participants who live in Bahri, Omdurman, and Khartoum use water for domestic purposes. In contrast, most of those who live in other areas use water for domestic and agricultural purposes. Most of the Bahri, Omdurman, and Khartoum participants and none of their relatives suffer from a waterborne disease. On the other hand, most of those who live in other areas or relatives suffer from a waterborne disease but with statistically no significant differences between those areas.

Table (35) showed that nil (color and PH Alkalinity) are nil in drinking water for all study areas.

Table (36) indicated that (87.5% and 86.7%) of water in Khartoum and Omdurman areas had nil Odor, while 100% of the water in Bahri was nil Odor.

The probability of the Pearson Chi-square (chi-square=0.446) was (Sig. =0.800), and the Likelihood Ratio (0.788) was (Sig. = 0.674), the both (Sig. values) were Greater than the alpha level of significance of 0.05. Therefore, the hypothesis that differences in the odor of water do not depend on the area. Thus all study areas have the same odor.

Table (37) revealed that (62.5% and 93.3%) of water in the Khartoum and Omdurman areas, respectively, have a clear appearance, while the water in Bahri has a 100% clear appearance.

The probability of the Pearson Chi-square statistics test (chi-square=0.4427) was (Sig. =0.109), and the Likelihood Ratio statistics test (0.4392) was (Sig. =0.111), the both (Sig. values) were greater than the alpha level of significance of 0.05. Therefore, the hypothesis that differences in the appearance of water do not become dependent on area. Thus all study areas have the same appearance.

Table (38) clearly determined that the (Asymp. Sig. values) for all parameters were greater than the alpha level of significance of 0.05 except Fluoride, Sulfate, and Potassium.

Therefore, all study areas have the same parameters other than (Fluoride, Sulfate, and Potassium).

2. Analysis of the results of water samples

The samples were collected from different sources of water as follows: The surface water sources such as Blue Nile (treated from taps and untreated directly from the Nile), White Nile (treated and untreated), Main Nile (treated and untreated), Groundwater, and wells water from Khartoum State: shajarah, kalakla, AlLamaab, JabraYathrub, AlOuzozab, Jabra, and Al Thourah.

The study concentrated on conducting a chemical analysis on samples of water taken from wells in some areas near to the industrial regions of the capital city Khartoum, e.g., HayAl Nuzha, Jabra, Al-Shagara, Al-Azouzab, Al-Lamb, Al-Kalakla, and Yathrib, while other samples were taken from the town of Omdurman, e.g., Al-Shati and Bant, etc. The different samples were taken from the city of Bahri, e.g., ElhagYousif, and El-Shaglah.

Parameters of chemical and physical analysis were determined, such as pH, Electric conductivity, suspended solids, Temperature, Turbidity, E-conductivity meter. Total dissolved solids (TDS) were measured by Direct reading using Standard (method NO 2130). Absorption Spectrophotometer (Model 1100) and the titration for some ions, and for $SO_4{-2}$ Cl-, $HCO_3{-2}$ and $CO_3{-2}$ were analyzed using a Dionex ion, chromatograph 2000i.

Some chemical characteristics were carried out for parameters of drinking water samples. These ionic components included: chlorine, Magnesium, Calcium, Chloride, Free Iron, total hardness, Total alkalinity, Nitrate, Nitrite using the methods such as (Quantitative inorganic chemistry), Ammonia(Nessler method), Hydrogen sulfide & sulfides, Sulfate, Iron ferrous:(1,10 phenanthroline method), and Fluoride, (SPANDS method)[7].

---

[7] Ref: standard method for the examination of water & wastewater.

To emphasize the characteristics of some ionic components:

1. **Chlorine:** Chlorine levels up to 5 milligrams per liter are considered safe in drinking water. At this level, harmful health effects are unlikely to occur (NHMRC & NRMMC, 2011).

2. **Magnesium and Calcium:** magnesium is responsible for causing most of the hardness and scale-forming properties of water. On the other hand, water low in calcium and magnesium is preferred in electroplating, tanning, dyeing, and textile manufacturing (Weight, 2008).

3. **Chloride:** Chloride ions in drinking water do not lead to harmful effects on health. However, if the levels are so high, then this can cause a salty taste. Chlorides do generally not damage people; but, if chloride is connected to the sodium part of table salt, this might harm the kidney and be linked to heart diseases (Omer N. H., 2019).

4. **Sodium:** According to (WHO, Sodium in Drinking-water, 2003), the total daily intake of sodium for growing infants and young children is 120–400 mg. On the other hand, adults require 500 mg of sodium to meet the daily needs. Regarding Sodium levels in drinking water, most water sources include less than 20 mg of sodium per liter; however, in some countries, levels can surpass 250 mg/liter. Also, when sodium levels reach levels above 200 mg/liter, then the water's taste will be changed. Based on (WHO, Sodium in Drinking-water, 2003), there has not been any conclusion concerning the potential relationship between sodium in drinking water and the occurrence of

hypertension. No health-based guideline value is therefore proposed.

5. **Iron and Manganese**: when there are elevated concentrations of these contaminants, this can lead to changing the color of water and a metallic taste to drinking water (Land, 1999). According to (Swistock, Sharpe, & Robillard, 2019), it is recommended that the levels of drinking water should not be more than 0.3 mg/L of iron and less than 0.05 mg/L of manganese. In case of having exceeded levels of iron in drinking water, several conditions can take place consequently. For example, it may allow bacteria to grow as iron acts as a host for bacteria. Moreover, there will be higher chances of getting acne and other skin conditions when drinking water has elevated iron content levels because iron closes pores and harm the skin cells. Furthermore, iron is stored in organs, such as the heart, pancreas, and liver. As a result, when the iron is being stored at a toxic level, it can cause iron poisoning, which is mainly damage to those organs. Iron poisoning symptoms include fatigue, weakness, joint pain, abdominal pain, heart or liver failure, and potentially lead to diabetes and loss of sex drive (BEdwards, 2019).

6. **Fluoride**: As discussed by (Demelash, Beyene, Abebe, & Melese, 2019), fluoride is contained in drinking water at low concentrations. Although fluoride is very crucial for teeth development, excessive exposure (more significant than the WHO guideline value of 1.5 mg/l) can possess several side effects.

7. **Nitrate**: The maximum contaminant level (MCL) for nitrate in public drinking water supplies in the United

States (U.S.) is ten mg/L as nitrate-nitrogen (NO3-N). This concentration is approximately equivalent to the World Health Organization (WHO) guideline of 50 mg/L as NO3 (Ward et al., 2018).

8. **Ammonia**: If ammonia is present in water at a level that exceeds the normal level, this can lead to ammonia poisoning. Drinking water that contains a high ammonia concentration for a long time can cause impairment to the human body's organ systems. The symptoms of ammonia poisoning include coughing, chest pain, fever, dizziness, nausea, vomiting, confusion, shock, collapse, and fainting. However, these symptoms do not usually occur because, in drinking water, the normal levels of ammonia are not high enough to cause them (Multipure, 2020).

The results of the analysis of all samples were compared with WHO standards in Khartoum state except for some data that were as follow.

The odor test showed positive results indicating that water is polluted. Ammonia and undissolved iron concentrations are more than the WHO's standard levels demonstrated in samples from Al Shajra, Al Azozab, and Al Kalaklah due to the presence of Al-Yarmook Factory near those wells. The best solution is to change its location to another position, alternatively to close the wells. All Omdurman samples' data agreed with the WHO standards with WHO except parameters such as TDS, Sodium, Chloride, and ammonia concentrations were more than the WHO's standard levels in Abu Seid and Al Thowra Hara 52. The reason behind this problem is that the soil is polluted. Therefore, pipes from unreactive metals should be used.

In Bahri Alhag Yosif Street 1: the Iron, undissolved Iron, and Chloride concentrations are higher than the WHO standard. Turbidity in individual wells is more than WHO standards, such as in AlAzozab, Jabra, Abu said, and Oumbada AlAmerya due to soil pollution and speculation of rainwater, as shown below:

1. Elshagra has a turbidity value of 1, a positive odor, a value of 1.6 of Ammonia, a value of Iron of 0.6, and an undissolved iron value is 0.66.mg/l

2. Alazozab has a turbidity value of 7 NTC

3. Eldobaseen has a turbidity value of 2.2 and a positive odor.

4. Al Kalakla has an Ammonia value of 1.57mg/l

- The exceptions were found in Omdurman: Abu Seid, Banat, Ombada Amreya, and Elthora Alhambra 52.

5. Abu Seid, one location had a TDS value of 1.469, chloride value of 270, ammonia value of 1.92, sodium value of 276, magnesium value of 0.035, and another location had a turbidity value of 8.4 and Un.dissolved iron value of 0.86.

6. Banat has a turbidity value of 7.

7. Ombada Amreya has a turbidity value of 10.4, and the nitrate value was 6

8. Elthora Alhamra 52 has a turbidity value of 6.6.

- The exceptions were found in Bahri: Elshaglah and Elhag Yousif St.1

9. Elshaglah has a temperature value of 30.9, an alkalinity value of 176, a value of 0.08 for Ammonia.

10. Elhag Yousif St.1 has a chloride value of 209, a value of 0.5 for Iron, and an undissolved iron value of 2.5.

According to the WHO (WHO, WATER QUALITY AND HEALTH - REVIEW OF TURBIDITY: Information for regulators, 2017), there were some incidents in which turbidity has been correlated with multiple diseases. Furthermore, the WHO has revealed that there have been studies that showed a relationship between turbidity and endemic disease, but others have not. Overall, it has not been confirmed that removal of turbidity can decrease the number of pathogens, so a constant relationship has not been established.

5.2. Conclusion

This study clearly showed that social media is responsible for informing the public in Khartoum State about Chemical water pollution. Still, their knowledge about water pollution is moderate. The source of drinking water for Khartoum residents is the Nile and the groundwater.

Most of the study area residents used water from wells, colorless with no odor or taste in Bhri, Omdurman, and Khartoum. Still, there was an odor in other residential areas. Water cuts were prevalent in the entire study area

- Most of the study people do not boil water before use, but some purify water by filtration.

- Some of the study area residents are water for drinking only white, others for drinking and agricultural purposes.

- By and large, the study area inhabitants are not suffering from water-borne or water-related diseases.

In the entire study, color, PH, and PH alkalinity are nil in the water, which had a clear appearance.

In the whole study area, water's chemical parameters are almost the same except Fluoride, Sulphate, and Potassium.

- All in all, in the study areas, the odors test was positive, indicating that water was polluted by Ammonia.

Figure 2: Location of polluted well drinking water

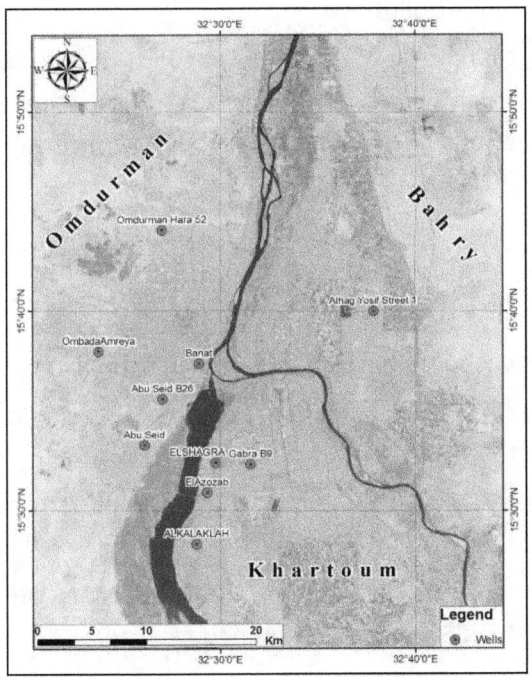

5.3. Recommendations

To improve the quality of water in Sudan, the study recommends:

1. Increase the awareness of the community about the effects of chemical water pollution by:
   a. Mass media
   b. Workshops, conferences, and seminars about chemical water pollution.
   c. Distributing posters and leaflets about chemical water pollution

2. The Government should invest in the water industry to be provided with healthy water, such as using a chemical treatment to reduce ammonia concentration and dissolved iron.

3. All citizens should be provided with safe water requirements with the WHO standards.

4. Use Isotopes Scopes in the entire labs to analyze and purify water in Khartoum State and other States.

5. Use solar energy for operating machines in stations and labs in the Khartoum States and other States.

6. In each residential block, there should be a network of sewage treatment.

7. Sudan should implement the millennium Development Goal (M.D.GS), which Sudan had ratified.

8. Periodical analysis for drinking water wells is critical.

9. Conduct scientific research on the quantities and qualities of groundwater in Khartoum state.

10. Construct dams to store rains water to be used after treatment in the rural areas in Khartoum state

References

New Jersey Administrative Code. (2020, October 5). Retrieved from Casetext: https://casetext.com/regulation/new-jersey-administrative-code/title-7-environmental-protection/chapter-9c-ground-water-quality-standards/subchapter-1-ground-water-quality-standards/section-79c-14-definitions

Abdeen, O. (2017). Demand for Biogas: State of the Art and Future Prospective. Omer, Abdeen, Demand for Biogas: State of the Art and Future Prospective (August 11,SSRN.

Abdellah, A. M., Abdel-Magi, H. M., & Yahia, N. A. (2012). Effect of Long-term Pumping on Fluoride Concentration Levels in Groundwater: A Case Study from East of Blue Nile Communities of Sudan. Journal of Applied Sciences, 1345-1354.

Abdellah, A. M., Abdel-Magid, H. M., & Shommo, E. I. (2013). Assessment of Groundwater Quality in Southern Suburb of the Omdurman City of Sudan. Greener Journal of Environmental Management and Public Safety, 83-90.

Abdel-Magid, H. M., Yahia, N. A., & Abdellah, A. M. (2012). Prediction of nitrate contamination trends of groundwater in Al-Butana region of Sudan. Environmental Science and Water Resources, 133 -143.

Ahmed, A. M., Sulaiman, W. N., Osman, M. M., Saeed, E. M., & Mohamed, Y. A. (2000). GROUNDWATER QUALITY IN KHARTOUM STATE, SUDAN. Journal of Environmental Hydrology.

Bartram, J., & Ballance, R. (1996). Water Quality Monitoring - A Practical Guide to the Design and Implementation of Freshwater Quality Studies and Monitoring Programmes. United Nations Environment Programme and the World Health Organization.

Beckedorf, A.-S. (2012). Political Waters: Governmental Water Management and Neoliberal Reforms in Khartoum/Sudan. Bayreuth: LIT Verlag Münster.

BEdwards. (2019, September 21). The Harmful Health Effects of Having Iron in Your Drinking Water. Retrieved from Hague Quality Water of

Maryland: https://www.haguewaterofmd.com/the-harmful-health-effects-of-having-iron-in-your-drinking-water/

Code, N. J. (2016, November 30). WATER QUALITY MANAGEMENT PLANNING Statutory authority: N.J.S.A. Retrieved from Lawinsider: https://www.lawinsider.com/documents/5QSnK0F2UCu

Committee, T. I. (2002). 2nd World Water Congress: Environmental Monitoring, Contaminants and Pathogens. Berlin.

D. Stephenson, E. S. (2004). Water Resources of Arid Areas: Proceedings of the International Conference on Water Resources of Arid and Semi-Arid Regions of Africa, Gaborone, Botswana. Taylor & Francis.

DANIELOPOL, D. L., GRIEBLER, C., GUNATILAKA, A., & GUNATILAKA, J. (2003). Present state and future prospects for groundwater ecosystems. Environmental Conservation, 104-130.

Demelash, H., Beyene, A., Abebe, Z., & Melese, A. (2019). Fluoride concentration in ground water and prevalence of dental fluorosis in Ethiopian Rift Valley: systematic review and meta-analysis. BMC Public Health.

Ell-Amin, A. M., Sulieman, A. E., & El-Khalifa, E. A. (2010). QUALITY CHARACTERISTICS OF DRINKING WATER IN KHARTOUM STATE AND WAD-MEDANI DISTRICT, SUDAN. Fourteenth International Water Technology Conference. Cairo.

Haga O. Mohamed, G. S. (2013). Introducing and Implementing an EMS in Khartoum State Water Corporation. Nile Basin Water Science & Engineering Journal.

Land, B. (1999). Iron and Manganese In Drinking Water. United States Department of Agriculture: Forest Service.

Multipure. (2020, March 31). Ammonia in Tap Water - Causes and Removal. Retrieved from Multipure: https://www.multipure.com/purely-social/science/ammonia-in-water/

Mohamed, H. O., & Skerratt, G. (2013). Introducing and Implementing an EMS in Khartoum State Water Corporation. Nile Basin Water Science & Engineering Journal.

NATIONS, F. A. (1995). Irrigation in Africa in figures. Rome.

NHMRC, & NRMMC. (2011). Australian Drinking Water Guidelines Paper 6 National Water Quality Management Strategy. Canberra: National Healthhand Medical Research Council,National Resource Management Ministerial Council.

Omer, A. (2002). Focus on groundwater in Sudan. Environmental Geology, 972-976.

Omer, N. H. (2019). Water Quality Parameters. Water Quality - Science, Assessments and Policy.

Schmoll, O., Howard, G., Chilton, J., & Chorus, I. (2006). Protecting groundwater for health: Managing the quality of drinking-water sources. London: World Health Organization.

Swistock, B. R., Sharpe, W. E., & Robillard, P. D. (2019). Iron and Manganese in Private Water Systems. Penn State Extension.

Ward, M. H., Jones, R. R., Brender, J. D., de Kok, T. M., Weyer, P. J., Nolan, B. T., et al. (2018). Drinking Water Nitrate and Human Health: An Updated Review. International journal of environmental research and public health.

Weight, W. (2008). Hydrogeology Field Manual, Second Edition. The McGraw-Hill Companies, Inc.

WHO. (2001). Parameters of Water Quality : Interpretation and Standards. Environmental Protection Agency.

WHO. (2003). Emerging Issues in Water and Infectious Disease. Environmental Protection Agency.

WHO. (2003). Sodium in Drinking-water. Guidelines for drinking-water quality.

WHO. (2017). WATER QUALITY AND HEALTH - REVIEW OF TURBIDITY: Information for regulators.

Yang, K., LeJeune, J., Alsdorf, D., Lu, B., Shum, C. K., & Liang, S. (2012). Global Distribution of Outbreaks of Water-Associated Infectious Diseases. PLoS neglected tropical diseases.

Appendices

Figure 1: The Mean of parameters in study areas

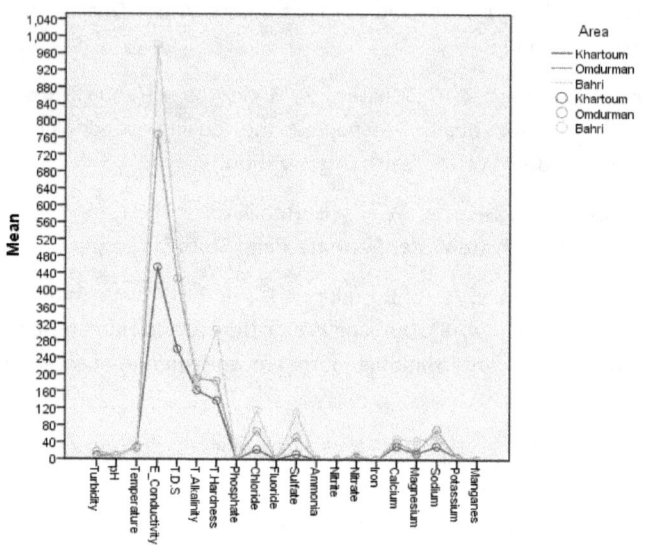

Questionnaire

Detection The Nile River Water and Drinking Wells Chemical Pollution in and Around Khartoum

I. Sociodemographic information

Q1. Age: _____

Q2. Sex: Female ( )   Male ( )

Q3. Where is your place of residence?

① Bahri

② Omdurman

③ Khartoum

④ Madani

⑤ Algadaref

⑥ Kasala

Other: _____

Q4. What is your highest level of education?

① Middle school or below

② High school

③ College or above

Q5. What is your marital status?

① Married

② Single

③ Widowed/Divorce

II. Level of Awareness

Please check the box that applies to you or elaborate in the case of "Others."

Q6. Have you ever heard about water pollution by chemicals?

① Yes

② No

③ Not sure

Q7. What was your source of information about water pollution by chemicals? Please check all that applies.

① Water pollution preventer

② Social media

③ Physician

④ Others:_____

Q8. If you would be informed of the water pollution by chemicals via one of the sources, how well do you believe you understand? (Circle the level of your understanding.)

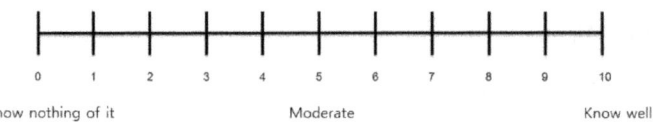

Q9. What is the source of your drinking water?

① Nile River

② Dug well

③ Drilled well

④ Not sure

Q10. What is the color of your drinking water?

① Whit

② Silver

③ Green

④ Yellow

⑤ colorless

⑥ other:_____

Q11. Does your drinking water have an odor?

① Yes

② No

③ Not sure

Q12. Does your drinking water have any taste?

① Yes

② No

③ Not sure

Q13. Do you experience drinking water cut?

① Yes

② No

③ Not sure

Q14. Do you boil water before use?

① Yes

② No

Q15. What do you use water for?

① Domestic uses only

② 1+ Agricultural uses

③ others :_____

Q16. Do you or one of your relatives suffer from a waterborne disease?

① Yes and the disease is :_____

② No

③ Not sure

Results

Location: 1

PARAMETER		UNIT	PARAMETER		UNIT
Appearance	Turbid		Chloride	6	Mg/l
Turbidity	10.5	NTU	Fluoride	0.2	Mg/l
Color	-	TCU	Sulfate	16	Mg/l
Odor	NIL		Ammonia	0.55	Mg/l
pH	7.9		Nitrite	0.034	Mg/l
Temperature	22.8	C°	Nitrate	7.1	Mg/l
E. Conductivity	234	Цs/cm	Iron	0.09	Mg/l
T.D.S	140	Mg/l	Calcium	24	Mg/l
T.S.S	-	Mg/l	Magnesium	8.64	Mg/l
T. Alkalinity	86	Mg/l	Sodium	7.8	Mg/l
PH.PH Alkalinity	NIL	Mg/l	Potassium	2.5	Mg/l
Hardness.T	96	Mg/l	Manganese	0.019	Mg/l
Phosphate	0.24	Mg/l			

All parameters within the permissible level of drinking water except turbidity.

Location: 2

PARAMETER		UNIT	PARAMETER		UNIT
Appearance	Turbid		Chloride	6	Mg/l
Turbidity	6.5	NTU	Fluoride	0.2	Mg/l
Color	-	TCU	Sulfate	16	Mg/l
Odor	NIL		Ammonia	0.31	Mg/l
pH	7.2		Nitrite	0.063	Mg/l
Temperature	23.1	C°	Nitrate	10.3	Mg/l
E. Conductivity	240	Цs/cm	Iron	0.01	Mg/l
T.D.S	143	Mg/l	Calcium	24	Mg/l
T.S.S	-	Mg/l	Magnesium	8.64	Mg/l
T. Alkalinity	86	Mg/l	Sodium	8.4	Mg/l
PH.PH Alkalinity	NIL	Mg/l	Potassium	3.0	Mg/l
Hardness.T	96	Mg/l	Manganese	0.015	Mg/l
Phosphate	0.2	Mg/l			

All parameters within the permissible level of drinking water except turbidity.

Location: 3

PARAMETER		UNIT	PARAMETER		UNIT
Appearance	Clear		Chloride	83	Mg/l
Turbidity	0.46	NTU	Fluoride	0.77	Mg/l
Color	-	TCU	Sulfate	42	Mg/l
Odor	NIL		Ammonia	0.02	Mg/l
pH	7.0		Nitrite	0.027	Mg/l
Temperature	29.5	C°	Nitrate	8.1	Mg/l
E. Conductivity	925	Цs/cm	Iron	0.05	Mg/l
T.D.S	508	Mg/l	Calcium	49.6	Mg/l
T.S.S	-	Mg/l	Magnesium	22.08	Mg/l
T. Alkalinity	222	Mg/l	Sodium	76	Mg/l
PH.PH Alkalinity	NIL	Mg/l	Potassium	5.8	Mg/l
Hardness.T	216	Mg/l	Manganese	0.013	Mg/l
Phosphate	0.41	Mg/l			

All parameters within the permissible level of drinking water

Location: 4

PARAMETER		UNIT	PARAMETER		UNIT
Appearance	Clear		Chloride	10	Mg/l
Turbidity	0.8	NTU	Fluoride	0.46	Mg/l
Color	-	TCU	Sulfate	6	Mg/l
Odor	NIL		Ammonia	NIL	Mg/l
pH	8.0		Nitrite	0.027	Mg/l
Temperature	27.4	C°	Nitrate	5.6	Mg/l
E. Conductivity	223	Цs/cm	Iron	0.08	Mg/l
T.D.S	134	Mg/l	Calcium	19.2	Mg/l
T.S.S	-	Mg/l	Magnesium	4.32	Mg/l
T. Alkalinity	84	Mg/l	Sodium	15	Mg/l
PH.PH Alkalinity	NIL	Mg/l	Potassium	6.2	Mg/l
Hardness.T	66	Mg/l	Manganese	0.007	Mg/l
Phosphate	0.21	Mg/l			

All parameters within the permissible level of drinking water except turbidity.

Location: 5

PARAMETER		UNIT	PARAMETER		UNIT
Appearance	Clear		Chloride	36	Mg/l
Turbidity	0.7	NTU	Fluoride	0.17	Mg/l
Color	-	TCU	Sulfate	7	Mg/l
Odor	NIL		Ammonia	0.1	Mg/l
pH	7.9		Nitrite	0.025	Mg/l
Temperature	23	C°	Nitrate	9.3	Mg/l
E. Conductivity	633	Цs/cm	Iron	0.06	Mg/l
T.D.S	348	Mg/l	Calcium	44	Mg/l
T.S.S	-	Mg/l	Magnesium	21.6	Mg/l
T. Alkalinity	230	Mg/l	Sodium	43.5	Mg/l
PH.PH Alkalinity	NIL	Mg/l	Potassium	12.8	Mg/l
Hardness.T	200	Mg/l	Manganese	0.014	Mg/l
Phosphate	0.4	Mg/l			

All parameters within the permissible level of drinking water

Location: 6

PARAMETER		UNIT	PARAMETER		UNIT
Appearance	Turbid		Chloride	6	Mg/l
Turbidity	29.1	NTU	Fluoride	0.1	Mg/l
Color	-	TCU	Sulfate	11	Mg/l
Odor	NIL		Ammonia	NIL	Mg/l
pH	7.1		Nitrite	0.028	Mg/l
Temperature	27.8	C°	Nitrate	5	Mg/l
E. Conductivity	228	Цs/cm	Iron	0.04	Mg/l
T.D.S	137	Mg/l	Calcium	25.6	Mg/l
T.S.S	-	Mg/l	Magnesium	8.16	Mg/l
T. Alkalinity	88	Mg/l	Sodium	8.1	Mg/l
PH.PH Alkalinity	NIL	Mg/l	Potassium	2.5	Mg/l
Hardness.T	98	Mg/l	Manganese	0.016	Mg/l
Phosphate	0.37	Mg/l			

All parameters within the permissible level of drinking water except turbidity.

Location: 7

PARAMETER		UNIT	PARAMETER		UNIT
Appearance	Clear		Chloride	12	Mg/l
Turbidity	1.5	NTU	Fluoride	0.27	Mg/l
Color	-	TCU	Sulfate	Nil	Mg/l
Odor	+ ve		Ammonia	1.6	Mg/l
pH	7.6		Nitrite	Nil	Mg/l
Temperature	31.3	C°	Nitrate	4.8	Mg/l
E. Conductivity	396	Ʉs/cm	Iron	0.06	Mg/l
T.D.S	223	Mg/l	Calcium	24	Mg/l
T.S.S	-	Mg/l	Magnesium	14.4	Mg/l
T. Alkalinity	154	Mg/l	Sodium	17.08	Mg/l
PH.PH Alkalinity	Nil	Mg/l	Potassium	5.98	Mg/l
Hardness.T	120	Mg/l	Manganese	-	Mg/l
Phosphate	0.03	Mg/l	Un. dissolved iron	0.66	

All parameters within the permissible level of drinking water except Ammonia and Un.dissolved iron

Location: 8

PARAMETER		UNIT	PARAMETER		UNIT
Appearance	Clear		Chloride	270	Mg/l
Turbidity	1.1	NTU	Fluoride	0.66	Mg/l
Color	-	TCU	Sulfate	245	Mg/l
Odor	NIL		Ammonia	1.92	Mg/l
pH	7.4		Nitrite	0.023	Mg/l
Temperature	29	C°	Nitrate	8.3	Mg/l
E. Conductivity	2672	Ʉs/cm	Iron	0.01	Mg/l
T.D.S	1469	Mg/l	Calcium	121.6	Mg/l
T.S.S	-	Mg/l	Magnesium	13.44	Mg/l
T. Alkalinity	342	Mg/l	Sodium	276	Mg/l
PH.PH Alkalinity	NIL	Mg/l	Potassium	11	Mg/l
Hardness.T	360*	Mg/l	Manganese	0.035	Mg/l

PARAMETER		UNIT			
Phosphate	0.26	Mg/l			

All parameters within the permissible level of drinking water except T.D.S, Chloride, Ammonia, and Sodium

Location: 9

PARAMETER		UNIT	PARAMETER		UNIT
Appearance	Clear		Chloride	16	Mg/l
Turbidity	0.71	NTU	Fluoride	0.16	Mg/l
Color	-	TCU	Sulfate	NIL	Mg/l
Odor	NIL		Ammonia	1.57	Mg/l
pH	7.7		Nitrite	0.033	Mg/l
Temperature	29.8	C°	Nitrate	2.0	Mg/l
E. Conductivity	473	Ⅱs/cm	Iron	0.04	Mg/l
T.D.S	284	Mg/l	Calcium	22.4	Mg/l
T.S.S	-	Mg/l	Magnesium	12.36	Mg/l
T. Alkalinity	200	Mg/l	Sodium	47.4	Mg/l
PH.PH Alkalinity	NIL	Mg/l	Potassium	4.2	Mg/l
Hardness.T	110	Mg/l	Manganese	0.153	Mg/l
Phosphate	0.49	Mg/l			

All parameters within the permissible level of drinking water except Ammonia

Location: 10

PARAMETER		UNIT	PARAMETER		UNIT
Appearance	Clear		Chloride	54	Mg/l
Turbidity	0.22	NTU	Fluoride	0.68	Mg/l
Color	-	TCU	Sulfate	39	Mg/l
Odor	NIL		Ammonia	0.01	Mg/l
pH	7.3		Nitrite	1.23	Mg/l
Temperature	28.7	C°	Nitrate	5.2	Mg/l
E. Conductivity	694.7	Ⅱs/cm	Iron	0.03	Mg/l
T.D.S	382.1	Mg/l	Calcium	35.2	Mg/l
T.S.S	-	Mg/l	Magnesium	20.64	Mg/l
T. Alkalinity	234	Mg/l	Sodium	76.94	Mg/l
PH.PH Alkalinity	NIL	Mg/l	Potassium	6.29	Mg/l

Hardness.T	174	Mg/l	Manganese	0.008	Mg/l
Phosphate	0.14	Mg/l			

All parameters within the permissible level of drinking water

Location: 11

PARAMETER		Unit
Appearance	Turbid	
Turbidity	8.4	NTU
Odor	NIL	
pH	7.5	
Tem.	28.9	C°
EC	470	ᒐs/cm
TDS	282	Mg/l
Chloride	19	Mg/l
T. Hardness	126	Mg/l
Ammonia	0.34	Mg/l
Iron	0.03	Mg/l
Un dissolve iron	0.86	Mg/l
Sodium	62.53	Mg/l

All parameters within the permissible level of drinking water except Turibidity and Undissolved iron

Location: 12

PARAMETER		UNIT	PARAMETER		UNIT
Appearance	Clear		Chloride	18	Mg/l
Turbidity	1.1	NTU	Fluoride	0.32	Mg/l
Color	-	TCU	Sulfate	10	Mg/l
Odor	NIL		Ammonia	0.01	Mg/l
pH	7.5		Nitrite	0.035	Mg/l
Temperature	22.2	C°	Nitrate	4.2	Mg/l
E. Conductivity	403	ᒐs/cm	Iron	0.01	Mg/l
T.D.S	201.5	Mg/l	Calcium	25.6	Mg/l
T.S.S	-	Mg/l	Magnesium	23.04	Mg/l
T. Alkalinity	160	Mg/l	Sodium	15.34	Mg/l
PH.PH Alkalinity	NIL	Mg/l	Potassium	2.63	Mg/l

PARAMETER		UNIT	PARAMETER		UNIT
Hardness.T	160	Mg/l	Un dissolve Iron	0.023	Mg/l
Phosphate	0.24	Mg/l			

All parameters within the permissible level of drinking water

Location: 13

PARAMETER		UNIT	PARAMETER		UNIT
Appearance	Clear		Chloride	290	Mg/l
Turbidity	2.66	NTU	Fluoride	0.56	Mg/l
Color	-	TCU	Sulfate	250	Mg/l
Odor	NIL		Ammonia	NIL	Mg/l
pH	7.3		Nitrite	0.028	Mg/l
Temperature	22.9	C°	Nitrate	24	Mg/l
E. Conductivity	1948.8	Ʋs/cm	Iron	0.5	Mg/l
T.D.S	974.4	Mg/l	Calcium	107.2	Mg/l
T.S.S	-	Mg/l	Magnesium	96	Mg/l
T. Alkalinity	178	Mg/l	Sodium	64.95	Mg/l
PH.PH Alkalinity	NIL	Mg/l	Potassium	2.45	Mg/l
Hardness.T	668*	Mg/l	Manganese	0.027	Mg/l
Phosphate	0.16	Mg/l	Un dissolve Iron	2.5	Mg/l

All parameters within the permissible level of drinking water except Chloride, iron, and undissolved iron

Location: 14

PARAMETER		UNIT	PARAMETER		UNIT
Appearance	Clear		Chloride	38	Mg/l
Turbidity	3.3	NTU	Fluoride	0.53	Mg/l
Color	-	TCU	Sulfate	64	Mg/l
Odor	NIL		Ammonia	-	Mg/l
pH	7.9		Nitrite	0.009	Mg/l
Temperature	28.3	C°	Nitrate	7.04	Mg/l
E. Conductivity	584	Ʋs/cm	Iron	0.08	Mg/l
T.D.S	292	Mg/l	Calcium	13.92	Mg/l
T.S.S	-	Mg/l	Magnesium	14.01	Mg/l
T. Alkalinity	175	Mg/l	Sodium	87.2	Mg/l

PH.PH Alkalinity	NIL	Mg/l	Potassium	1.54	Mg/l
Hardness.T	93.2	Mg/l	Manganese	-	Mg/l
Phosphate	0.22	Mg/l			

All parameters within the permissible level of drinking water

Location: 15

Sample		Unit
Appearance	Turbid	
Turbidity	7	NTU
Odor	NIL	
pH	7.2	
Tem.	28	° C
EC	498	Цs/cm
TDS	301	Mg/l
T. Alkalinity	198	Mg/l
Chloride	56	Mg/l
T. Harness	192	Mg/l
Calcium	41.6	Mg/l
Magnesium	21.12	Mg/l
Ammonia	NIL	Mg/l

All parameters within the permissible level of drinking water except Turbidity

Location: 16

Parameter	Sample	Unite
Appearance	Turbid	
Turbidity	6.6	NTU
Odor	NIL	
pH	6.8	
Tem.	30	° C
EC	509	µs/cm
TDS	356	mg/l
Iron	0.06	mg/l
Un dissolve Iron	0.24	mg/l

All parameters within the permissible level of drinking water except Turbidity

Location: 17

Parameter	Sample	Unite
Appearance	Clear	
Turbidity	4.59	NTU
Odor	NIL	
Tem.	27.5	° C
pH	7.5	
EC	1020	µs/cm
TDS	714	mg/l
Ammonia	0.65	mg/l
H.Sulfide	0.01	
Total Coliform	Zero	colony/100ml

All parameters within the permissible level of drinking water

Location: 18

PARAMETER		UNIT	PARAMETER		UNIT
Appearance	Clear		Chloride	74	Mg/l
Turbidity	4.4	NTU	Fluoride	0.25	Mg/l
Color	-	TCU	Sulfate	49	Mg/l
Odor	NIL		Ammonia	0.33	Mg/l
pH	7.8		Nitrite	0.268	Mg/l
Temperature	27.8	C°	Nitrate	4.0	Mg/l
E. Conductivity	824	Цs/cm	Iron	0.08	Mg/l
T.D.S	453.2	Mg/l	Calcium	46.4	Mg/l
T.S.S	-	Mg/l	Magnesium	24	Mg/l
T. Alkalinity	238	Mg/l	Sodium	81.3	Mg/l
PH.PH Alkalinity	NIL	Mg/l	Potassium	5.7	Mg/l
Hardness.T	216	Mg/l	Manganese	0.042	Mg/l
Phosphate	0.12	Mg/l	Un dissolve iron	0.86	Mg/l

All parameters within the permissible level of drinking water except Phosphate.

Location: 19

PARAMETER		UNIT	PARAMETER		UNIT
Appearance	Turbid		Chloride	64	Mg/l
Turbidity	10.4	NTU	Fluoride	0.43	Mg/l
Color	-	TCU	Sulfate	52	Mg/l
Odor	Nil		Ammonia	0.13	Mg/l
pH	7.9		Nitrite	-	Mg/l
Temperature	25.1	C°	Nitrate	6.0	Mg/l
E. Conductivity	826.8	Цs/cm	Iron	0.03	Mg/l
T.D.S	454.7	Mg/l	Calcium	43.2	Mg/l
T.S.S	-	Mg/l	Magnesium	21.12	Mg/l
T. Alkalinity	110	Mg/l	Sodium	81.36	Mg/l
PH.PH Alkalinity	Nil	Mg/l	Potassium	11.96	Mg/l
Hardness.T	196	Mg/l	Manganese	-	Mg/l
Phosphate	0.03	Mg/l			

The amount of turbidity exceeds the permissible level of drinking water. Others within limits.

Location: 20

Parameter	Sample	Unite
Turbidity	2.1	NTU
Odor	+ve	
pH	7.7	
Temperature	31.0	C°
E. Conductivity	472	Цs/cm
TDS	260	mg/l
Ammonia	0.53	mg/l
H.Sulfide	0.022*	mg/l
Total coliform	Zero	Colony / 100ml

All parameters within the permissible level of drinking water except there is an odor

Location: 21

Parameter	Sample	Unite
Turbidity	2.1	NTU

Odor	+ve	
pH	7.7	
Temperature	31.0	C°
E. Conductivity	472	Цs/cm
TDS	260	mg/l
Ammonia	0.53	mg/l
H.Sulfide	0.022*	mg/l
Total coliform	Zero	Colony / 100ml

All parameters within the permissible level of drinking water except there is an odor

Location: 22

Parameter	Sample	Unite
Color	-VE	
Odor	-VE	
Appearance	Clear	
Turbidity	0.77	NTU
Temperature	30.9	C°
EC	551	Цs/cm
TDS	331	mg/l
Alkalinity	176	mg/l
Ammonia	0.08	mg/l
H. Sulfide	0.002	mg/l

All parameters within the permissible level of drinking water

Location: 23

PARAMETER		UNIT	PARAMETER		UNIT
Appearance	Clear		Chloride	18	Mg/l
Turbidity	0.58	NTU	Fluoride	0.39	Mg/l
Color	-	TCU	Sulfate	18.1	Mg/l
Odor	+ve		Ammonia	0.185	Mg/l
pH	7.3		Nitrite	0.075	Mg/l
Temperature	27.9	C°	Nitrate	1.6	Mg/l
E. Conductivity	401	Цs/cm	Iron	0.05	Mg/l
T.D.S	221	Mg/l	Calcium	24	Mg/l
T.S.S	-	Mg/l	Magnesium	11.52	Mg/l

T. Alkalinity	144	Mg/l	Sodium	34.14	Mg/l
PH.PH Alkalinity	NIL	Mg/l	Potassium	8.44	Mg/l
Hardness.T	108	Mg/l	Manganese	0.022	Mg/l
Phosphate	0.13	Mg/l			

All parameters within the permissible level of drinking

WHO Guidelines and Sudanese Standards for Drinking Water

No	Component	WHO Guidelines value ppm	SSMO Standard ppm
1	EC	-	3000 (µs/cm)
2	TDS	1000	1000
3	pH	6.5-8.5	6.5-8.5 (pH unit)
4	Turbidity	5 NTU	6 NTU
5	T.H	500	500
6	Ca	-	240
7	Mg	-	125
8	AS	0.01	0.007
9	CD	0.003	0.003
10	K		10
11	Fe	0.3	0.3
12	F	1.5	1.5
13	Ag	0.1	-
14	Cr	0.05	0.04
15	Cu	2	1.5
16	Mn	0.5	0.4
17	Na	200	200
18	Al	0.2	0.2
19	Ba	0.7	0.7
20	Pb	0.01	0.007
21	Chloride	250	250
22	Chlorine	0.5	-
23	CN	0.07	-
24	Nitrate	50	50
25	Nitrite	3	2
26	Ammonia	1.5	1.5
27	Bicarbonate	-	500
28	Sulphate	400	250

WHO, 2006, and SSMO 2002[8].

---

[8] Ref: Standard method for the examination of water & waste water (14) edition

www.ingramcontent.com/pod-product-compliance
Lightning Source LLC
Chambersburg PA
CBHW072215170526
45158CB00002BA/614